Biology For Beginners 2025 Edition

The Comprehensive Step-By-Step Guide for Students to Effortlessly Understand and Master the Study of Living Things with Clarity and Enjoyment

Ethan D. Needleman

Copyright Notice

No part of this book may be reproduced, written, electronic, recorded, or photocopied without written permission from the publisher or author.

The exception would be in the case of brief quotations embodied in critical articles or reviews and pages where permission is specifically granted by the publisher or author.

Although every precaution has been taken to verify the accuracy of the information contained herein, the author and publisher assume no responsibility for any errors or omissions. No liability is assumed for damages that may result from the use of the information contained within.

All Rights Reserved ©2025

TABLE OF CONTENTS

INTRODUCTION   7

PURPOSE OF THE BOOK   7
How to Use This Book   8
Tips for Studying Biology   9

PART I: THE BASICS OF BIOLOGY   12

CHAPTER 1: WHAT IS BIOLOGY?   13
Definition and Scope   13
Importance of Biology in Daily Life   15
Branches of Biology   17
CHAPTER 2: THE SCIENTIFIC METHOD   21
Steps of the Scientific Method   21
Experiment Design   24
Analyzing Data and Drawing Conclusions   27
CHAPTER 3: BASIC CHEMISTRY FOR BIOLOGY   33
Atoms and Molecules   33
Chemical Bonds   36
Water and Its Properties   39
Acids, Bases, and pH   42

PART II: THE BUILDING BLOCKS OF LIFE   46

CHAPTER 4: CELL STRUCTURE AND FUNCTION   47
The Cell Theory   47
Prokaryotic vs. Eukaryotic Cells   50
Cell Organelles and Their Functions   54
Cell Membrane Structure and Function   59
CHAPTER 6: CELLULAR PROCESSES   63
Diffusion and Osmosis   63
Cellular Respiration   66
Photosynthesis   69
Cell Cycle and Mitosis   72
Meiosis and Sexual Reproduction   75

PART III: GENETICS   78

CHAPTER 7: DNA AND RNA   79
Structure of DNA and RNA   79

DNA Replication	**82**
Transcription and Translation	**86**
Chapter 8: Mendelian Genetics	**91**
Gregor Mendel's Experiments	91
Laws of Inheritance	94
Punnett Squares	97
Dominant and Recessive Traits	101
Chapter 9: Modern Genetics	**105**
Genetic Mutations	105
Genetic Engineering and Biotechnology	107
Ethical Considerations in Genetics	110

PART IV: EVOLUTION AND DIVERSITY — 114

Chapter 10: Principles of Evolution	**115**
Theories of Evolution	115
Natural Selection	118
Speciation	122
Chapter 11: Classification of Living Organisms	**127**
The Five Kingdoms	127
The Three Domains of Life	131
Taxonomy and Phylogeny	134
Protists	138
Fungi	140
Viruses and Prions	143

PART V: THE PLANT KINGDOM — 148

Chapter 13: Plant Structure and Function	**149**
Plant Cells and Tissues	149
Roots, Stems, and Leaves	151
Plant Transport Systems	155
Chapter 14: Plant Physiology	**159**
Photosynthesis in Detail	159
Plant Growth and Development	162
Reproduction in Plants	166
Chapter 15: Introduction to Animals	**171**
Animal Cells and Tissues	171
Animal Body Plans and Symmetry	175
Worms and Mollusks	178
Arthropods and Echinoderms	182
Chapter 17: Vertebrates	**187**
Fish, Amphibians, and Reptiles	187

Birds and Mammals 191

PART VII: ECOLOGY AND THE ENVIRONMENT 196

Chapter 18: Principles of Ecology 197
Ecosystems and Biomes 197
Population Dynamics 201
Chapter 19: Interactions in the Ecosystem 205
Symbiosis and Competition 205
Predator-Prey Relationships 208
Human Impact on Ecosystems 211
Conservation Strategies 215
Sustainable Practices 221
Homeostasis and Regulation 228
Chapter 22: Nutrition and Digestion 233
Nutrients and Their Functions 233
The Digestive System 237
Healthy Eating and Metabolism 241
Chapter 23: Circulatory and Respiratory Systems 247
Structure and Function of the Heart and Blood Vessels 247
Blood and Lymphatic System 252
The Respiratory System and Gas Exchange 255
Chapter 24: The Nervous and Endocrine Systems 259
Neurons and Nerve Impulses 259
The Brain and Spinal Cord 262
Hormones and Glands 265
Fertilization and Embryonic Development 269
Growth and Aging 273
Chapter 26: Conclusion 277
Glossary of Terms 277
Common Abbreviations in Biology 281

INTRODUCTION

Ever felt overwhelmed by the complexities of biology? Are you struggling to connect the dots between abstract concepts and real-life applications? Do you find it challenging to grasp the basics of cell structure or the intricacies of genetics? Have you ever wondered how microorganisms impact our world or how the principles of evolution shape the diversity of life?

If these questions resonate with you, you're not alone. Many students and learners alike face hurdles when diving into the vast and intricate world of biology. The subject can seem daunting, with its endless terminologies, detailed processes, and ever-evolving discoveries. But what if learning biology didn't have to be so intimidating? What if you could approach this fascinating science with clarity, confidence, and even enjoyment?

Welcome to **Biology For Beginners 2025: The Comprehensive Step-By-Step Guide for Students to Effortlessly Understand and Master the Study of Living Things with Clarity and Enjoyment**. This book is designed to transform your understanding of biology from confusing to captivating. Whether you're a student embarking on your first biology course or someone looking to refresh your knowledge, this guide offers a clear, engaging, and structured approach to mastering biology.

Biology For Beginners 2025 breaks down complex concepts into manageable sections, starting with the fundamentals and gradually advancing to more intricate topics. From the basics of cell structure and genetics to the wonders of evolution and ecology, each chapter is crafted to build your knowledge step-by-step. With clear explanations, helpful tips, and real-world connections, this book aims to make biology not just understandable but enjoyable.

Prepare to embark on a journey where biology becomes not just a subject to study, but a fascinating exploration of life itself. With this guide in hand, you'll gain a deeper appreciation for the living world around you and develop a solid foundation in biological sciences. Let's make your learning experience in biology both enlightening and exciting!

Purpose of the Book

Biology For Beginners 2025: The Comprehensive Step-By-Step Guide for Students to Effortlessly Understand and Master the Study of Living Things with Clarity and Enjoyment is crafted with one central goal: to make the study of biology accessible, engaging, and rewarding for everyone. Here's how this book aims to fulfill that purpose:

Simplify Complex Concepts: Biology is a vast and intricate field, but our approach is designed to break down complicated ideas into easy-to-understand sections. By presenting information in a clear, step-by-step manner, we help you grasp foundational concepts without feeling overwhelmed.

Build a Solid Foundation: Starting from the basics, this book provides a structured progression through essential topics. Whether you're new to biology or need a refresher, each chapter builds on the last, ensuring a strong understanding of fundamental principles before moving on to more advanced subjects.

Engage and Inspire: Learning biology doesn't have to be a chore. With engaging explanations, practical examples, and real-world connections, we aim to spark curiosity and foster a genuine interest in the study of living things.

Provide Practical Study Tips: Effective study strategies are crucial for mastering any subject. This book offers practical tips and techniques for studying biology efficiently, helping you retain information and apply it confidently.

Support Diverse Learning Styles: We recognize that everyone learns differently. To accommodate various learning preferences, the book includes visual aids, summaries, and review questions that cater to different styles, making the material more accessible.

Highlight Real-Life Relevance: Biology is not just an academic subject but a science that impacts our daily lives. This book emphasizes how biological concepts apply to everyday experiences and current issues, helping you see the relevance and importance of what you're learning.

Prepare for Future Studies: For those planning to advance in biological sciences or related fields, this book lays a strong groundwork. By mastering the basics and understanding key principles, you'll be well-prepared for more advanced study and specialized topics.

In essence, **Biology For Beginners 2025** is more than just a textbook; it's a guide designed to make your journey through the world of biology both clear and enjoyable. With this book, you'll gain a deeper appreciation for the living world and a solid foundation in biological science.

How to Use This Book

Biology For Beginners 2025 is designed to be your comprehensive guide to mastering the fundamentals of biology. To get the most out of this book, follow these steps:

Start with the Basics: Begin by reading the introductory sections, including the "Purpose of the Book" and "Tips for Studying Biology." These sections will help you understand the book's structure and provide strategies to enhance your learning experience.

Follow the Structure: The book is divided into distinct parts and chapters, each covering specific topics. Progress through the sections in order, as the material is organized to build your knowledge progressively. Starting with the basics in Part I and advancing through subsequent sections will ensure a solid foundation before tackling more complex subjects.

Utilize the Key Features:

- **Definitions and Key Terms**: Pay attention to highlighted terms and definitions. They are crucial for understanding the concepts discussed in each chapter.
- **Diagrams and Illustrations**: Visual aids are provided to help clarify complex concepts. Refer to these diagrams often to reinforce your understanding.
- **Summary Sections**: Each chapter concludes with a summary of key points. Review these summaries to consolidate your learning and check your comprehension.

Engage with Review Questions: At the end of each chapter, you'll find review questions and exercises. Use these to test your understanding and reinforce the material. Answering these questions will help solidify your knowledge and prepare you for more advanced topics.

Apply What You Learn: Wherever possible, connect the concepts you're learning to real-world examples and applications. Understanding how biological principles relate to everyday life can enhance your comprehension and retention of the material.

Refer to the Glossary: For any terms or concepts that are unclear, consult the glossary at the end of the book. This will help clarify definitions and reinforce your understanding of specific terms.

Review Regularly: Periodically review previous chapters and summaries to keep concepts fresh in your mind. Regular revision will help maintain your grasp of the material as you progress through the book.

Seek Additional Resources: If you find a particular topic challenging, don't hesitate to seek additional resources such as online tutorials, videos, or supplementary texts. This can provide further clarification and different perspectives on the material.

By following these guidelines, you'll be able to navigate **Biology For Beginners 2025** effectively, making your study of biology clear, structured, and enjoyable. Dive in with curiosity and an open mind, and let this book be your guide to mastering the fascinating world of living things.

Tips for Studying Biology

Studying biology can be both exciting and challenging. To help you make the most of your learning experience, here are some effective tips and strategies:

Understand the Basics First: Before diving into complex topics, ensure you have a solid grasp of fundamental concepts. Understanding basic principles, such as cell structure and the scientific method, will make learning advanced topics much easier.

Use Active Learning Techniques: Engage with the material actively rather than passively reading. Try techniques such as summarizing information in your own words, teaching concepts to someone else, or creating mind maps to visualize relationships between ideas.

Employ Visual Aids: Biology is a highly visual science with many structures and processes. Use diagrams, charts, and videos to help you understand and remember key concepts. Drawing your own diagrams can also reinforce your understanding.

Make Flashcards: Create flashcards for key terms, definitions, and processes. Flashcards are great for quick reviews and can help reinforce memory through active recall and spaced repetition.

Relate Concepts to Real Life: Connect what you're learning to real-world examples. Understanding how biological principles apply to everyday life or current scientific issues can make the material more relevant and memorable.

Practice with Questions: Regularly work through practice questions and problems. This not only tests your understanding but also helps you apply concepts in different contexts. Review and analyze your answers to identify areas for improvement.

Study Regularly and Consistently: Set up a regular study schedule to avoid cramming. Consistent, shorter study sessions are more effective than last-minute intensive study sessions. Regular review of material helps reinforce learning and improves retention.

Join Study Groups: Studying with peers can provide new insights and make learning more interactive. Group discussions can help clarify difficult concepts and expose you to different perspectives.

Utilize Online Resources: Take advantage of online resources such as educational videos, interactive simulations, and reputable websites. These can provide additional explanations and examples to complement your textbook.

Stay Organized: Keep your notes, textbooks, and study materials well-organized. Use folders or digital tools to track your progress, important concepts, and areas where you need further review.

Take Breaks and Stay Healthy: Avoid burnout by taking regular breaks during study sessions. Ensure you get enough sleep, eat well, and exercise regularly. A healthy body supports a healthy mind, enhancing your ability to learn and retain information.

Ask for Help When Needed: Don't hesitate to seek help if you're struggling with a concept. Reach out to teachers, tutors, or online forums for clarification and additional support.

By applying these tips, you'll be well-equipped to tackle the challenges of studying biology and gain a deeper understanding of this fascinating field.

Part I: The Basics of Biology

Chapter 1: What is Biology?

Definition and Scope

Definition: Biology is the scientific study of life and living organisms. It explores the structure, function, growth, origin, evolution, and distribution of living organisms. The term "biology" is derived from the Greek words "bios," meaning life, and "logos," meaning study or discourse. Thus, biology is essentially the study of life in all its forms and complexities.

Scope of Biology: The scope of biology is vast and encompasses numerous sub-disciplines and aspects. Here's a detailed look at various components:

Levels of Biological Organization:

Molecular Biology: This level focuses on the chemical processes within and related to living organisms. Molecular biology is concerned with the structure and function of macromolecules, such as DNA, RNA, and proteins, and how these molecules interact to govern life processes.

Cell Biology: This branch studies cells, the basic units of life. It includes understanding cell structure, function, and processes such as metabolism, cell division, and cellular communication.

Organismal Biology: This area examines individual organisms, including their anatomy, physiology, development, and behavior. It can be further divided into various subfields such as botany (study of plants), zoology (study of animals), and microbiology (study of microorganisms).

Ecology: This field explores the interactions between organisms and their environments. It includes studying ecosystems, biodiversity, population dynamics, and the impact of environmental changes on living organisms.

Evolutionary Biology: This branch focuses on the origin and changes in the diversity of life over time. It includes studying the processes of natural selection, genetic drift, and speciation that contribute to the evolution of species.

Applications of Biology:

Medical and Health Sciences: Biology is crucial in medicine, including understanding diseases, developing treatments, and advancing medical technologies. Fields like genetics, immunology, and pharmacology are directly tied to biological research.

Agriculture: Biology plays a key role in improving crop yields, developing pest-resistant plants, and understanding soil health. Agricultural biology involves studying plant and animal breeding, genetics, and sustainable farming practices.

Environmental Science: Biology helps address environmental issues by studying ecosystems, conservation, and the impact of human activities on the environment. This includes efforts to preserve biodiversity and manage natural resources.

Methodologies in Biology:

Experimental Biology: Involves conducting experiments to test hypotheses and understand biological processes. This includes laboratory work, field studies, and data analysis.

Comparative Biology: Compares biological systems across different organisms to understand commonalities and differences. This approach can reveal evolutionary relationships and functional adaptations.

Systems Biology: Focuses on understanding complex interactions within biological systems. This includes using computational models to study networks of genes, proteins, and other molecules.

Integration with Other Disciplines:

Biochemistry: Combines biology and chemistry to explore the chemical processes within living organisms.

Biophysics: Applies principles of physics to understand biological systems, such as the mechanics of molecular interactions.

Bioinformatics: Uses computational tools to analyze biological data, particularly large datasets from genomics and proteomics.

Key Concepts in Biology:

Cell Theory: The concept that all living organisms are composed of cells, and cells are the basic unit of life.

Gene Theory: The idea that genes are the units of heredity and that they determine the traits of organisms.

Homeostasis: The ability of organisms to maintain a stable internal environment despite external changes.

Evolution: The process through which species undergo genetic changes over time, leading to diversity and adaptation.

Importance of Biology:

Understanding Life Processes: Biology helps us understand how living organisms function, adapt, and evolve, providing insights into the nature of life itself.

Solving Real-World Problems: Biological research addresses critical issues such as health care, environmental conservation, and food security, impacting society and the planet.

In summary, biology is a comprehensive and dynamic field that covers a wide range of topics related to life and living organisms. Its scope extends from the molecular to the ecological level, integrating knowledge from various disciplines and contributing to numerous applications that impact our world.

Importance of Biology in Daily Life

Biology impacts nearly every aspect of our daily lives, from the food we eat to the healthcare we receive. Here's a detailed examination of its relevance:

1. Health and Medicine

Understanding Diseases:

Diagnosis and Treatment: Biology helps in diagnosing diseases by understanding their underlying biological mechanisms. For example, knowledge of microbiology helps in identifying pathogens, while genetics aids in understanding inherited conditions.

Vaccines and Antibiotics: The development of vaccines and antibiotics relies on biological research. Vaccines stimulate the immune system to protect against diseases, while antibiotics target bacterial infections.

Personal Health:

Genetic Testing: Advances in genetics allow for personalized medicine, where treatments can be tailored to an individual's genetic makeup. Genetic testing can also help in assessing the risk of inherited diseases.

Nutrition: Understanding the biology of nutrition and metabolism helps in making informed dietary choices, leading to better health outcomes.

2. Agriculture and Food Production

Crop Improvement:

Genetic Engineering: Biology plays a crucial role in improving crop yields and resistance to pests through genetic modification. This includes creating genetically modified organisms (GMOs) that can thrive in various conditions.

Sustainable Practices: Biological research contributes to sustainable agricultural practices by studying soil health, pest control, and crop rotation.

Food Safety:

Microbiology: Knowledge of microbiology helps in ensuring food safety by preventing contamination and spoilage. It involves understanding how bacteria and other microorganisms affect food.

3. Environmental Conservation

Ecosystem Management:

Biodiversity: Biology helps in understanding ecosystems and the importance of biodiversity. Protecting various species and their habitats is crucial for maintaining ecological balance.

Climate Change: Biological research helps in studying the effects of climate change on ecosystems and species, providing insights into mitigation and adaptation strategies.

Pollution Control:

Bioremediation: Biology is used in cleaning up polluted environments through bioremediation, where living organisms break down pollutants in soil and water.

4. Everyday Products and Technologies

Biotechnology:

Bioprocessing: Biotechnology applies biological processes to produce products like biofuels, pharmaceuticals, and enzymes used in various industries.

Medical Devices: Advances in biology contribute to the development of medical devices and technologies, such as diagnostic tools and prosthetics.

Consumer Products:

Biopolymers: Biological research contributes to the development of biopolymers, which are environmentally friendly alternatives to conventional plastics.

5. Education and Awareness

Scientific Literacy:

Informed Decisions: Understanding basic biological concepts helps individuals make informed decisions about health, environment, and technology.

Critical Thinking: Biology encourages critical thinking and scientific inquiry, which are valuable skills in various aspects of life.

6. Reproduction and Genetics

Family Planning:

Reproductive Health: Biology provides insights into reproductive health and family planning, including understanding fertility, contraception, and prenatal care.

Genetic Inheritance:

Inheritance Patterns: Knowledge of genetics helps in understanding inheritance patterns and genetic disorders, which can be important for family planning and managing genetic conditions.

7. Understanding Life Processes

Body Functioning:

Homeostasis: Biology explains how the body maintains internal stability (homeostasis) despite external changes. This includes understanding processes like temperature regulation, hydration, and metabolism.

Aging: Biological research into aging processes helps in understanding how aging affects the body and in developing strategies to promote healthy aging.

8. Social and Ethical Issues

Biotechnology Ethics:

Ethical Considerations: Advances in biotechnology and genetic engineering raise ethical questions about the use of these technologies. Understanding biology helps in navigating these ethical issues and making informed decisions.

Public Policy:

Policy Making: Biological research informs public policies related to health, environment, and agriculture. Knowledge of biology can help in advocating for policies that support sustainable and equitable practices.

Biology is integral to many aspects of daily life, from health and medicine to agriculture, environmental conservation, and technology. Its applications touch on personal well-being, societal issues, and global challenges. By understanding biology, individuals can make better decisions, contribute to solving pressing issues, and appreciate the complexity of the natural world. The importance of biology extends beyond the classroom, influencing everyday activities and shaping the future of our planet and species.

Branches of Biology

1. Molecular Biology

Focus:

Molecular biology studies the molecular mechanisms that govern the structure and function of genes and proteins. It is concerned with the interactions between different systems of a cell, including the interactions between DNA, RNA, and proteins.

Key Concepts:

Genetics: Involves the study of genes, genetic variation, and heredity. Molecular biology and genetics are closely related, as molecular biology explores how genetic information is transferred and expressed.

Protein Synthesis: Examines how cells produce proteins based on genetic instructions, including transcription (DNA to RNA) and translation (RNA to protein).

Gene Regulation: Studies how gene expression is controlled and how it affects cellular function and development.

2. Cell Biology

Focus:

Cell biology investigates the structure, function, and behavior of cells, which are the basic units of life. It includes the study of cell organelles, cell metabolism, cell division, and cellular communication.

Key Concepts:

Cell Structure: Involves understanding organelles such as the nucleus, mitochondria, endoplasmic reticulum, and Golgi apparatus.

Cell Cycle: Studies the processes of cell growth, DNA replication, and cell division (mitosis and meiosis).

Cell Signaling: Examines how cells communicate with each other through signaling molecules and pathways.

3. Genetics

Focus:

Genetics explores heredity, genetic variation, and the roles of genes in determining traits and susceptibility to diseases. It encompasses the study of gene function, inheritance patterns, and genetic mutations.

Key Concepts:

Mendelian Genetics: Involves the principles of inheritance established by Gregor Mendel, including dominant and recessive traits.

Genomics: Studies the entire genome of an organism, including gene mapping and sequencing.

Epigenetics: Investigates changes in gene expression that do not involve alterations to the DNA sequence itself.

4. Developmental Biology

Focus:

Developmental biology examines the process by which organisms grow and develop from fertilization to maturity. It includes studying embryonic development, morphogenesis, and differentiation.

Key Concepts:

Embryogenesis: The process of development from a fertilized egg to a fully formed organism.

Cell Differentiation: How unspecialized cells become specialized to perform specific functions.

Morphogenesis: The development of the structure and form of an organism.

5. Evolutionary Biology

Focus:

Evolutionary biology explores the origins and changes in the diversity of life over time. It studies the processes that drive evolutionary change and the relationships between different organisms.

Key Concepts:

Natural Selection: The mechanism proposed by Charles Darwin that explains how adaptive traits become more common in a population.

Speciation: The process by which new species arise from existing ones.

Phylogenetics: The study of evolutionary relationships among organisms, often depicted in phylogenetic trees.

6. Ecology

Focus:

Ecology is the study of interactions between organisms and their environments. It includes understanding ecosystems, population dynamics, and the impact of human activities on the environment.

Key Concepts:

Ecosystems: Complex networks of interactions between living organisms and their physical environment.

Biomes: Large ecological areas characterized by their climate, vegetation, and animal life.

Conservation Biology: Focuses on preserving biodiversity and managing natural resources.

7. Physiology

Focus:

Physiology examines how the body's organs and systems function and how they contribute to maintaining homeostasis. It covers both plant and animal physiology.

Key Concepts:

Homeostasis: The mechanisms that organisms use to maintain stable internal conditions.

Organ Systems: Study of specific systems such as the circulatory, respiratory, digestive, and nervous systems in animals.

Plant Physiology: Involves understanding how plants function, including processes like photosynthesis and nutrient uptake.

8. Microbiology

Focus:

Microbiology is the study of microorganisms, including bacteria, viruses, fungi, and protozoa. It explores their physiology, genetics, and roles in health, disease, and the environment.

Key Concepts:

Pathogen Biology: Studies how microorganisms cause diseases and how they interact with hosts.

Microbial Ecology: Examines the roles of microorganisms in ecosystems, including their impact on nutrient cycles.

Industrial Microbiology: Involves the use of microorganisms in industrial processes, such as fermentation and biotechnology.

9. Botany

Focus:

Botany is the study of plants, including their structure, function, growth, and taxonomy. It encompasses a wide range of plant-related topics.

Key Concepts:

Plant Anatomy: The study of the internal structure of plants.

Plant Ecology: Focuses on plant interactions with their environment and other organisms.

Plant Evolution: Examines the evolutionary history of plant species and their adaptations.

10. Zoology

Focus:

Zoology is the study of animals, including their physiology, behavior, and classification. It covers a wide range of animal life from invertebrates to vertebrates.

Key Concepts:

Animal Behavior: The study of how animals interact with each other and their environment.

Animal Anatomy: Focuses on the structure of animals and how it relates to their function.

Evolutionary Zoology: Examines the evolutionary relationships and history of animal species.

Conclusion

The branches of biology each focus on different aspects of living organisms and their interactions with the environment. By studying these various branches, scientists and researchers can gain a comprehensive understanding of life, from the molecular and cellular levels to ecosystems and evolutionary processes. Each branch contributes to a broader knowledge base that is essential for addressing biological questions, solving real-world problems, and advancing scientific and medical research.

Chapter 2: The Scientific Method
Steps of the Scientific Method

The scientific method is a structured approach used to conduct research and acquire knowledge in a systematic and objective manner. It involves a series of steps designed to ensure that investigations are thorough, unbiased, and replicable. Here's a detailed exploration of each step:

1. Observation

Definition:

Observation is the initial step where a researcher identifies a phenomenon or problem that sparks curiosity. This can be anything from noticing a pattern, a peculiar occurrence, or an unanswered question in the natural world.

Examples:

Observing that plants in a particular area seem to grow faster than plants in another area.

Noticing that a certain medication appears to improve symptoms in patients more effectively than others.

Importance:

Observations help to generate questions and hypotheses and provide the foundation for further investigation.

2. Question

Definition:

Based on the observations, a specific question is formulated that the research aims to answer. The question should be clear, focused, and researchable.

Examples:

Why do plants in one area grow faster than those in another area?

Does the medication actually have a higher efficacy compared to other treatments?

Importance:

A well-defined question directs the research process and helps in formulating hypotheses.

3. Hypothesis

Definition:

A hypothesis is a tentative explanation or prediction that can be tested through experimentation or further observation. It is often framed as an "if... then..." statement.

Examples:

If plants are given a specific type of fertilizer, then they will grow faster than those given other types of fertilizer.

If the medication contains a particular active ingredient, then it will show improved efficacy in treating the symptoms.

Importance:

Hypotheses provide a basis for experimental design and allow researchers to test their predictions in a controlled manner.

4. Experiment

Definition:

Experiments are designed to test the validity of the hypothesis through controlled procedures. This involves manipulating variables and observing the outcomes.

Key Components:

Variables:

Independent Variable: The factor that is deliberately changed or manipulated in the experiment (e.g., type of fertilizer).

Dependent Variable: The factor that is measured or observed in response to changes in the independent variable (e.g., plant growth rate).

Control Variables: Factors that are kept constant to ensure that any changes in the dependent variable are due to the manipulation of the independent variable.

Controls: Experimental conditions that ensure the experiment is fair and valid. This can include control groups that do not receive the experimental treatment.

Examples:

Conducting an experiment where one group of plants receives a specific fertilizer, another group receives a different type, and a control group receives no fertilizer. Measuring and comparing growth rates across these groups.

Importance:

Experiments provide empirical evidence to support or refute the hypothesis and contribute to a deeper understanding of the observed phenomenon.

5. Data Collection and Analysis

Definition:

Data collection involves gathering and recording information from the experiment. Analysis involves interpreting this data to determine whether it supports or contradicts the hypothesis.

Methods:

Quantitative Data: Numerical data that can be measured and analyzed statistically (e.g., growth measurements, chemical concentrations).

Qualitative Data: Descriptive data that can provide insights into patterns and observations (e.g., plant appearance, behavioral observations).

Importance:

Accurate data collection and analysis are crucial for drawing valid conclusions and ensuring the reliability of the research findings.

6. Conclusion

Definition:

The conclusion is drawn based on the analysis of the data. It involves evaluating whether the results support the hypothesis or if the hypothesis needs to be revised.

Examples:

Concluding that the specific fertilizer did result in faster plant growth based on the experimental data.

Determining that the medication did not show improved efficacy compared to other treatments.

Importance:

The conclusion helps in summarizing the findings and determining the next steps, which may include further experimentation or research.

7. Communication

Definition:

Communicating the results involves sharing the findings with the scientific community and the public through reports, publications, presentations, or discussions.

Methods:

Scientific Papers: Detailed reports of the research process, methods, results, and conclusions.

Conferences: Presentations and discussions with other scientists and researchers.

Publications: Articles or papers published in scientific journals or magazines.

Importance:

Effective communication allows others to review, replicate, and build upon the research. It also contributes to the advancement of knowledge and scientific progress.

8. Revision and Further Research

Definition:

Based on feedback, new observations, or additional questions, the research process may be revisited. This may involve refining hypotheses, designing new experiments, or exploring related topics.

Importance:

Science is an iterative process where findings lead to new questions and further investigation. Continuous revision and research help in refining theories and expanding knowledge.

The scientific method is a fundamental process that ensures rigorous and objective inquiry in science. By following the steps of observation, questioning, hypothesizing, experimenting, analyzing, concluding, communicating, and revising, researchers can systematically explore phenomena, validate findings, and contribute to the advancement of scientific knowledge. This methodical approach helps in minimizing bias, improving the reliability of results, and fostering a deeper understanding of the natural world.

Experiment Design

Experiment design is a crucial aspect of the scientific method, as it involves planning and executing a study to test a hypothesis in a systematic and controlled manner. Proper experiment design ensures that results are reliable, valid, and can be replicated. Here's an extensive discussion on the key components and considerations in designing an experiment:

Experiment Design

1. Formulating a Hypothesis

Definition:

A hypothesis is a testable prediction or explanation for a phenomenon. It is often phrased as an "if... then..." statement that provides a clear and measurable expectation.

Example:

Hypothesis: "If plants are exposed to blue light, then they will grow taller than plants exposed to red light."

Importance:

A well-formulated hypothesis provides a focused objective for the experiment and guides the design of the study.

2. Identifying Variables

Definition:

Variables are factors that can affect the outcome of an experiment. Properly identifying and controlling variables is essential for ensuring valid results.

Types of Variables:

Independent Variable: The factor that is deliberately changed or manipulated by the researcher to test its effect (e.g., type of light in the plant growth experiment).

Dependent Variable: The factor that is measured or observed in response to changes in the independent variable (e.g., plant height).

Control Variables: Factors that are kept constant to ensure that any changes in the dependent variable are due to the manipulation of the independent variable (e.g., soil type, water amount, and temperature).

Importance:

Properly identifying and managing variables ensures that the experiment tests the hypothesis effectively and that results are not influenced by external factors.

3. Designing the Experimental Procedure

Steps in Designing the Procedure:

Step 1: Outline the steps involved in conducting the experiment, including how the independent variable will be manipulated and how the dependent variable will be measured.

Step 2: Determine the sample size, or the number of subjects or samples to be used. A larger sample size can provide more reliable results.

Step 3: Decide on the method of data collection and ensure that it is consistent and accurate.

Example:

Procedure: Place one set of plants under blue light and another set under red light for a specified period. Measure the height of the plants at regular intervals.

Importance:

A well-defined procedure ensures that the experiment can be conducted consistently and that the data collected will be relevant to the hypothesis.

4. Control Group vs. Experimental Group

Definitions:

Control Group: A group that is not exposed to the experimental treatment and is used for comparison. It helps to establish a baseline against which the effects of the independent variable can be measured.

Experimental Group: The group that is exposed to the experimental treatment or condition.

Example:

In a plant growth experiment, the control group might receive normal light, while the experimental group receives blue light.

Importance:

Having a control group allows researchers to determine if the observed effects are due to the manipulation of the independent variable or other factors.

5. Randomization

Definition:

Randomization involves randomly assigning subjects or samples to different groups to reduce bias and ensure that each group is comparable.

Methods:

Random Sampling: Selecting subjects randomly from a larger population.

Random Assignment: Randomly assigning subjects to different experimental or control groups.

Importance:

Randomization helps to ensure that the results are not skewed by pre-existing differences between groups, leading to more reliable conclusions.

6. Replication

Definition:

Replication involves repeating the experiment to verify the results and ensure that they are consistent and reliable.

Types:

Within-Experiment Replication: Repeating the same experiment under similar conditions.

Between-Experiment Replication: Conducting similar experiments with different conditions or variables to see if the results hold true across different settings.

Importance:

Replication strengthens the validity of the findings and helps to confirm that the results are not due to chance.

7. Data Collection and Analysis

Data Collection:

Methods: Collect data systematically using tools and techniques appropriate for the type of data being gathered (e.g., measurements, observations, surveys).

Accuracy: Ensure that data collection methods are accurate and consistent to avoid errors.

Data Analysis:

Statistical Analysis: Use statistical methods to analyze the data and determine if the results support the hypothesis. This may include calculations of averages, standard deviations, and significance tests.

Interpretation: Analyze the data to draw conclusions about the hypothesis, considering whether the observed effects are significant and if they support the initial prediction.

Importance:

Proper data collection and analysis are crucial for drawing valid conclusions and determining the significance of the results.

8. Reporting Results

Definition:

Reporting involves presenting the findings of the experiment in a clear and comprehensive manner. This may include writing a report, presenting at a conference, or publishing in a scientific journal.

Components:

Introduction: Background information, hypothesis, and objectives of the experiment.

Methods: Detailed description of the experimental design and procedures.

Results: Presentation of data in tables, graphs, and figures.

Discussion: Interpretation of results, including how they relate to the hypothesis and previous research.

Conclusion: Summary of findings and implications.

Importance:

Effective reporting allows others to understand, evaluate, and build upon the research. It contributes to the scientific body of knowledge and facilitates peer review and replication.

Experiment design is a fundamental component of the scientific method that involves careful planning and execution to test hypotheses and answer research questions. By meticulously designing experiments, identifying and controlling variables, using appropriate methods, and analyzing results, researchers can ensure that their findings are reliable and meaningful. Good experimental design not only advances scientific knowledge but also contributes to the credibility and reproducibility of research findings.

Analyzing Data and Drawing Conclusions

Analyzing data and drawing conclusions are critical steps in the scientific method that follow the experimental phase. These steps are crucial for interpreting the results of an experiment and determining whether they support the hypothesis. Here's an extensive discussion on these aspects:

Analyzing Data

1. Data Collection

Definition:

Data collection is the process of gathering information during an experiment. This information can be quantitative (numerical) or qualitative (descriptive).

Methods:

Quantitative Data: Measured and recorded in numerical terms (e.g., temperature readings, plant growth measurements).

Qualitative Data: Observational and descriptive (e.g., color changes, behavioral patterns).

Importance:

Accurate and systematic data collection ensures that the analysis will be based on reliable and relevant information.

2. Organizing Data

Definition:

Organizing data involves arranging it in a structured format to facilitate analysis. This can include creating tables, charts, and graphs.

Methods:

Tables: Display data in rows and columns for easy comparison and reference.

Graphs: Visual representations of data, such as bar graphs, line graphs, and scatter plots, which help to illustrate trends and relationships.

Charts: Pie charts or histograms that show distributions and proportions.

Importance:

Proper organization helps in identifying patterns, trends, and outliers, making it easier to analyze the data effectively.

3. Statistical Analysis

Definition:

Statistical analysis involves applying mathematical techniques to interpret data and determine whether the observed effects are significant.

Key Concepts:

Descriptive Statistics: Summarize and describe the main features of the data, including measures of central tendency (mean, median, mode) and dispersion (range, variance, standard deviation).

Inferential Statistics: Used to make inferences or predictions about a population based on sample data. This includes hypothesis testing, confidence intervals, and p-values.

Methods:

Hypothesis Testing: Determines if there is enough evidence to reject the null hypothesis. Common tests include t-tests, chi-square tests, and ANOVA (Analysis of Variance).

Correlation and Regression Analysis: Examines the relationship between variables and predicts the value of one variable based on another.

Importance:

Statistical analysis helps to determine if the results are due to chance or if they reflect a true effect, thus validating the findings of the experiment.

4. Interpreting Results

Definition:

Interpreting results involves making sense of the data analysis and understanding what the findings mean in the context of the hypothesis and research questions.

Considerations:

Patterns and Trends: Identify any consistent patterns or trends in the data.

Significance: Assess whether the results are statistically significant and if they support or refute the hypothesis.

Comparison: Compare results with previous studies to understand how they fit into the broader body of knowledge.

Importance:

Accurate interpretation ensures that conclusions are based on a thorough understanding of the data and its implications.

Drawing Conclusions

1. Evaluating Hypothesis

Definition:

Drawing conclusions involves evaluating whether the experimental results support or refute the hypothesis.

Steps:

Support for Hypothesis: If the data shows a clear effect as predicted, the hypothesis is supported. However, it's essential to ensure that the findings are statistically significant.

Refutation of Hypothesis: If the data does not show the expected effect or shows contrary results, the hypothesis may be refuted or revised.

Importance:

Drawing accurate conclusions helps in validating the hypothesis or refining it for further research.

2. Considering Experimental Limitations

Definition:

Limitations are potential weaknesses or constraints in the experimental design that could affect the validity of the results.

Types:

Experimental Errors: Errors in measurement, data collection, or procedures that could influence the results.

Sample Size: A small sample size may not provide a representative result or may increase variability.

External Factors: Variables not controlled in the experiment that could impact the outcomes.

Importance:

Acknowledging limitations helps in understanding the context and reliability of the findings and guides future research directions.

3. Drawing Implications

Definition:

Drawing implications involves interpreting how the results contribute to the broader field of study and what they suggest for practical applications.

Considerations:

Theoretical Implications: How the results fit into existing theories and models.

Practical Implications: Potential applications of the findings in real-world settings, such as in medicine, technology, or policy.

Importance:

Understanding the implications of the results provides insights into their significance and relevance beyond the specific experiment.

4. Communicating Findings

Definition:

Communicating findings involves presenting the results and conclusions to the scientific community and the public through various means.

Methods:

Scientific Papers: Detailed reports published in academic journals that include an introduction, methods, results, discussion, and conclusion.

Presentations: Sharing results at conferences or seminars.

Reports and Summaries: Providing summaries and reports for non-specialist audiences or stakeholders.

Importance:

Effective communication ensures that the research contributes to the collective knowledge and allows others to review, critique, and build upon the findings.

5. Revising and Further Research

Definition:

Based on the conclusions, further research may be needed to address unanswered questions, refine the hypothesis, or explore new areas of inquiry.

Considerations:

Follow-Up Experiments: Design additional experiments to verify results or investigate new questions.

Research Gaps: Identify areas where further research could provide additional insights or resolve uncertainties.

Importance:

Continuous research and revision help to advance knowledge, refine scientific theories, and address emerging questions or issues.

Analyzing data and drawing conclusions are pivotal stages in the scientific method that involve careful examination of experimental results to determine their significance and implications. Through systematic data collection, statistical analysis, and thoughtful interpretation, researchers can validate hypotheses, identify patterns, and contribute to the broader scientific understanding. Effective communication and acknowledgment of experimental limitations further enhance the credibility and impact of the research.

Chapter 3: Basic Chemistry for Biology
Atoms and Molecules

1. Atoms

Definition:

Atoms are the basic units of matter and the fundamental building blocks of all substances. Each atom consists of a nucleus surrounded by electrons.

Components of Atoms:

Nucleus: The central part of the atom, containing protons and neutrons.

Protons: Positively charged particles found in the nucleus. The number of protons defines the element and is known as the atomic number.

Neutrons: Neutral particles (with no charge) in the nucleus that contribute to the atomic mass but do not affect the chemical properties of the element.

Electrons: Negatively charged particles that orbit the nucleus in electron shells or energy levels. Electrons are involved in chemical bonding and reactions.

Atomic Structure:

Electron Shells: Electrons are arranged in concentric shells around the nucleus. Each shell can hold a specific number of electrons.

First Shell: Can hold up to 2 electrons.

Second Shell: Can hold up to 8 electrons.

Third Shell: Can hold up to 18 electrons, and so on.

Valence Electrons: Electrons in the outermost shell are called valence electrons. They play a key role in chemical bonding and reactions.

Atomic Number and Mass Number:

Atomic Number: The number of protons in the nucleus, which defines the element (e.g., carbon has an atomic number of 6).

Mass Number: The total number of protons and neutrons in the nucleus. It determines the atomic mass of the element.

Isotopes:

Atoms of the same element with different numbers of neutrons are called isotopes. Isotopes have the same chemical properties but different atomic masses (e.g., Carbon-12 and Carbon-14).

2. Molecules

Definition:

Molecules are formed when two or more atoms bond together through chemical bonds. They represent the smallest unit of a chemical compound that can participate in a chemical reaction.

Types of Molecules:

Diatomic Molecules: Molecules consisting of two atoms (e.g., O_2, N_2).

Polyatomic Molecules: Molecules consisting of three or more atoms (e.g., H_2O, CO_2).

Chemical Bonds:

Covalent Bonds: Formed when two atoms share one or more pairs of electrons. Covalent bonds can be single, double, or triple, depending on the number of shared electron pairs.

Polar Covalent Bonds: Occur when the sharing of electrons between atoms is unequal, resulting in partial positive and negative charges (e.g., H_2O).

Nonpolar Covalent Bonds: Occur when electrons are shared equally between atoms (e.g., O_2).

Ionic Bonds: Formed when one atom donates electrons to another, creating ions with opposite charges that attract each other (e.g., NaCl).

Hydrogen Bonds: Weak interactions between a hydrogen atom covalently bonded to an electronegative atom and another electronegative atom (e.g., the bonds between water molecules).

Molecular Formulas:

Represent the number and type of atoms in a molecule (e.g., H_2O indicates two hydrogen atoms and one oxygen atom).

Structural Formulas:

Show the arrangement of atoms within a molecule and the bonds between them. Examples include Lewis structures and Kekulé structures.

3. Chemical Reactions

Definition:

Chemical reactions involve the breaking and forming of chemical bonds to transform reactants into products. These reactions are fundamental to biological processes.

Types of Reactions:

Synthesis (Combination) Reactions: Two or more substances combine to form a more complex substance (e.g., A + B → AB).

Decomposition Reactions: A complex substance breaks down into simpler substances (e.g., AB → A + B).

Single Replacement Reactions: One element replaces another in a compound (e.g., AB + C → AC + B).

Double Replacement Reactions: Exchange of ions between two compounds (e.g., AB + CD → AD + CB).

Redox Reactions: Involve the transfer of electrons between substances, changing their oxidation states (e.g., cellular respiration).

Reaction Rates:

Factors Affecting Reaction Rates: Concentration, temperature, pressure, and the presence of catalysts can influence the rate of chemical reactions.

Equilibrium:

Dynamic Equilibrium: In reversible reactions, the rates of the forward and reverse reactions are equal, resulting in stable concentrations of reactants and products.

4. Importance in Biology

Biological Molecules:

Water (H_2O): Essential for life, water has unique properties such as high heat capacity, cohesion, and adhesion, which are crucial for maintaining biological functions.

Organic Molecules: Carbon-based molecules, including carbohydrates, lipids, proteins, and nucleic acids, are fundamental to biological structures and functions.

Carbohydrates: Provide energy and structural support (e.g., glucose, starch, cellulose).

Lipids: Involved in energy storage, membrane structure, and signaling (e.g., fats, oils, phospholipids).

Proteins: Serve as enzymes, structural components, and signaling molecules (e.g., enzymes, antibodies).

Nucleic Acids: Store and transmit genetic information (e.g., DNA, RNA).

Chemical Reactions in Metabolism:

Catabolic Reactions: Break down molecules to release energy (e.g., cellular respiration).

Anabolic Reactions: Build complex molecules from simpler ones, requiring energy input (e.g., protein synthesis).

Understanding atoms and molecules is fundamental to the study of biology because they are the building blocks of matter and the basis for biochemical processes. Atoms combine to form molecules through various types of chemical bonds, and these molecules participate in chemical reactions essential for life. By studying the principles of chemistry, including atomic structure, bonding, and reactions, we gain insights into the molecular mechanisms underlying biological functions and processes.

Chemical Bonds

1. Ionic Bonds

Definition:

Ionic bonds are formed through the electrostatic attraction between oppositely charged ions. This type of bond occurs when one atom donates electrons to another, creating positive and negative ions.

Formation:

Electron Transfer: An atom with a few electrons in its outer shell (typically a metal) loses one or more electrons, becoming a positively charged cation. An atom with nearly full outer shells (typically a nonmetal) gains these electrons, becoming a negatively charged anion.

Electrostatic Attraction: The resulting cations and anions attract each other due to their opposite charges, forming an ionic bond.

Example:

Sodium chloride (NaCl) is a classic example where sodium (Na) donates an electron to chlorine (Cl). Sodium becomes a Na$^+$ ion, and chlorine becomes a Cl$^-$ ion. The electrostatic attraction between these ions forms the ionic bond.

Properties:

High Melting and Boiling Points: Ionic compounds typically have high melting and boiling points due to the strong electrostatic forces holding the ions together.

Solubility in Water: Many ionic compounds are soluble in water, as water molecules can surround and separate the ions.

Electrical Conductivity: Ionic compounds conduct electricity when dissolved in water or melted, as the ions are free to move and carry an electrical current.

2. Covalent Bonds

Definition:

Covalent bonds are formed when two atoms share one or more pairs of electrons to achieve full outer electron shells and stabilize themselves.

Types of Covalent Bonds:

Single Covalent Bond: Involves the sharing of one pair of electrons between two atoms (e.g., H_2, where each hydrogen atom shares one electron).

Double Covalent Bond: Involves the sharing of two pairs of electrons between two atoms (e.g., O_2, where each oxygen atom shares two pairs of electrons).

Triple Covalent Bond: Involves the sharing of three pairs of electrons between two atoms (e.g., N_2, where each nitrogen atom shares three pairs of electrons).

Polarity:

Nonpolar Covalent Bonds: Occur when electrons are shared equally between two atoms of the same or similar electronegativity (e.g., O_2, N_2). The electron density is evenly distributed.

Polar Covalent Bonds: Occur when electrons are shared unequally between atoms with different electronegativities, leading to partial positive and negative charges (e.g., H_2O). The molecule has a dipole moment.

Properties:

Lower Melting and Boiling Points: Covalent compounds often have lower melting and boiling points compared to ionic compounds.

Solubility: Many covalent compounds are not soluble in water but may dissolve in organic solvents.

Electrical Insulation: Covalent compounds generally do not conduct electricity, as they lack free-moving charged particles.

3. Metallic Bonds

Definition:

Metallic bonds are formed between metal atoms where electrons are shared freely among a lattice of positive metal ions. This "sea of electrons" is responsible for many of the physical properties of metals.

Formation:

Electron Delocalization: In metallic bonding, electrons are not bound to any specific atom but move freely throughout the metal lattice, creating a strong bond between metal ions.

Properties:

Electrical Conductivity: Metals are good conductors of electricity due to the free movement of electrons.

Malleability and Ductility: Metals can be hammered into thin sheets (malleability) or drawn into wires (ductility) because the metallic bonds can be easily reformed without breaking.

Luster: Metals have a shiny appearance due to the reflection of light off the free electrons.

4. Hydrogen Bonds

Definition:

Hydrogen bonds are weak interactions that occur between a hydrogen atom covalently bonded to a more electronegative atom and another electronegative atom.

Formation:

Hydrogen Atom Interaction: In a molecule with a hydrogen atom bonded to a highly electronegative atom (e.g., oxygen or nitrogen), the partial positive charge on the hydrogen atom attracts the partial negative charge of another electronegative atom in a nearby molecule.

Examples:

Water Molecules: In water (H_2O), hydrogen bonds form between the hydrogen atoms of one water molecule and the oxygen atoms of adjacent water molecules.

DNA: Hydrogen bonds stabilize the double helix structure of DNA by connecting complementary bases (adenine with thymine, and guanine with cytosine).

Properties:

Relatively Weak: Hydrogen bonds are weaker than covalent and ionic bonds but are crucial for the stability and function of many biological molecules.

Influence on Properties: Hydrogen bonding contributes to the high boiling point of water, its surface tension, and its solvent properties.

5. Van der Waals Forces

Definition:

Van der Waals forces are weak, short-range forces that arise from temporary dipoles created when electrons are unevenly distributed in atoms or molecules.

Types:

London Dispersion Forces: Occur between all molecules due to temporary fluctuations in electron density, leading to temporary dipoles.

Dipole-Dipole Interactions: Occur between molecules with permanent dipoles, where the positive end of one dipole attracts the negative end of another.

Properties:

Weak and Temporary: Van der Waals forces are weaker than hydrogen bonds and are significant only when molecules are very close together.

Role in Biological Systems: They contribute to the overall interactions between molecules and are important for protein folding and molecular recognition.

Importance of Chemical Bonds in Biology

Biological Molecules:

Proteins: Proteins are composed of amino acids linked by peptide bonds (a type of covalent bond). The structure and function of proteins are determined by various types of bonds, including hydrogen bonds and disulfide bonds.

Nucleic Acids: DNA and RNA contain covalent bonds between nucleotides and hydrogen bonds between base pairs. These bonds are crucial for genetic information storage and transfer.

Carbohydrates: Carbohydrates consist of sugar molecules connected by glycosidic bonds, which are covalent bonds that link monosaccharides into complex polysaccharides.

Chemical Reactions:

Enzymes: Enzymes are proteins that catalyze chemical reactions by lowering the activation energy. Their function depends on the precise bonding and interactions between substrates and active sites.

Metabolism: Metabolic processes involve breaking and forming chemical bonds to convert energy and build or degrade molecules. Understanding these bonds helps explain how energy is transferred and utilized in biological systems.

Chemical bonds are fundamental to the structure and function of molecules, which are essential for all biological processes. Ionic, covalent, metallic, and hydrogen bonds each play distinct roles in determining the properties and interactions of molecules. Understanding these bonds provides insight into the molecular basis of life, from the stability of biological macromolecules to the mechanisms of biochemical reactions.

Water and Its Properties

1. Structure of Water Molecules

Molecular Formula:

Water has the molecular formula H_2O, indicating that each water molecule consists of two hydrogen atoms covalently bonded to one oxygen atom.

Molecular Geometry:

Bent Shape: The water molecule has a bent or V-shaped structure due to the two lone pairs of electrons on the oxygen atom that push the hydrogen atoms closer together. The bond angle between the hydrogen-oxygen-hydrogen atoms is approximately 104.5°.

Polarity:

Polar Molecule: Water is a polar molecule with a partial negative charge on the oxygen atom and partial positive charges on the hydrogen atoms. This polarity arises because oxygen is more electronegative than hydrogen, causing an uneven distribution of electron density.

2. Unique Properties of Water

a. High Heat Capacity

Definition:

Heat capacity is the amount of heat required to change the temperature of a substance by a given amount. Water has a high heat capacity, meaning it can absorb or release a large amount of heat with only a small change in temperature.

Significance:

Thermal Regulation: Water's high heat capacity helps organisms maintain a stable internal temperature despite external temperature fluctuations. This property is crucial for homeostasis in living organisms.

Climate Regulation: Large bodies of water, such as oceans and lakes, moderate the Earth's climate by absorbing and releasing heat, influencing weather patterns and temperature.

b. High Heat of Vaporization

Definition:

Heat of vaporization is the amount of energy required to convert a liquid into a vapor. Water has a high heat of vaporization, meaning it requires a significant amount of energy to evaporate.

Significance:

Cooling Mechanism: The high heat of vaporization of water allows for effective cooling through processes such as sweating and transpiration. When water evaporates from the surface of the skin or plants, it absorbs heat, helping to cool the organism or environment.

Energy Exchange: The evaporation of water from oceans and lakes contributes to energy exchange in the atmosphere, influencing weather and climate patterns.

c. High Surface Tension

Definition:

Surface tension is the cohesive force between molecules at the surface of a liquid that causes the liquid to behave as if it has an elastic skin. Water has high surface tension due to the strong hydrogen bonds between water molecules.

Significance:

Capillary Action: Water's high surface tension enables capillary action, where water moves through narrow spaces against gravity. This property is essential for the transport of water and nutrients in plants.

Formation of Droplets: Surface tension causes water to form droplets, which is evident in the way water beads up on surfaces.

d. Cohesion and Adhesion

Definition:

Cohesion: The attraction between molecules of the same substance. In water, cohesion is due to hydrogen bonding between water molecules.

Adhesion: The attraction between molecules of different substances. Water's adhesion allows it to stick to various surfaces.

Significance:

Meniscus Formation: Cohesion and adhesion are responsible for the formation of a meniscus in a graduated cylinder, where water curves up at the edges due to adhesion to the glass.

Transport in Plants: Cohesion and adhesion are crucial for the movement of water through the plant vascular system, contributing to the process of transpiration and nutrient transport.

e. Universal Solvent

Definition:

Water is known as the universal solvent because it can dissolve a wide variety of substances, particularly those with ionic or polar properties.

Significance:

Chemical Reactions: Water's ability to dissolve many substances makes it a vital medium for biochemical reactions and processes within cells.

Nutrient Transport: In organisms, water dissolves nutrients, gases, and waste products, facilitating their transport and removal.

f. Density and Ice Formation

Definition:

Water is unique in that its solid form (ice) is less dense than its liquid form. This occurs because of the hydrogen bonding in ice, which forms a crystalline structure with more open spaces compared to liquid water.

Significance:

Insulation of Aquatic Life: Ice floats on water, providing an insulating layer that helps to protect aquatic life during cold temperatures. This insulation helps maintain a stable environment for organisms in lakes and oceans.

Climate Impact: The presence of ice caps and glaciers influences global climate and sea levels.

3. Biological Importance of Water

a. Biological Functions

Metabolic Reactions:

Water is involved in numerous metabolic reactions, including hydrolysis and condensation. Hydrolysis involves breaking down molecules with the addition of water, while condensation involves the formation of larger molecules with the removal of water.

Solvent for Biological Molecules:

Water dissolves essential biological molecules such as proteins, nucleic acids, and carbohydrates, facilitating their interactions and functions within cells.

b. Cellular Processes

Cell Structure:

Water provides turgor pressure in plant cells, helping to maintain cell shape and structural integrity. The pressure exerted by the fluid in the central vacuole of plant cells supports the cell wall.

Temperature Regulation:

Through processes such as sweating and evaporation, water helps regulate body temperature in animals and maintains thermal balance in organisms.

c. Homeostasis

Fluid Balance:

Water plays a key role in maintaining fluid balance within cells, tissues, and organs. Proper hydration is essential for physiological functions and overall health.

Transport of Substances:

Water serves as the medium for transporting nutrients, gases, and waste products throughout organisms, ensuring that cells receive necessary substances and eliminate by-products.

Water's unique properties, such as high heat capacity, high heat of vaporization, high surface tension, cohesion, adhesion, and its role as a universal solvent, make it indispensable to life. These properties influence many biological processes, including temperature regulation, nutrient transport, and cellular functions. Understanding water's behavior and significance helps explain its central role in supporting life and maintaining ecological balance.

Acids, Bases, and pH

1. Acids

Definition:

Acids are substances that can donate a proton (H^+ ion) to another substance. In aqueous solutions, acids increase the concentration of hydrogen ions (H^+).

Characteristics of Acids:

Taste: Acids generally have a sour taste (e.g., citric acid in lemons).

pH: Acids have a pH less than 7.

Reactivity: Acids react with metals to produce hydrogen gas and with bases to form salts and water.

Types of Acids:

Strong Acids: Fully dissociate into their ions in water, resulting in a high concentration of hydrogen ions. Examples include hydrochloric acid (HCl) and sulfuric acid (H_2SO_4).

Weak Acids: Partially dissociate in water, resulting in a lower concentration of hydrogen ions. Examples include acetic acid (CH_3COOH) and citric acid ($C_6H_8O_7$).

Examples in Biology:

Lactic Acid: Produced during anaerobic respiration in muscles.

Carbonic Acid: Formed in the blood from carbon dioxide and plays a role in maintaining blood pH.

2. Bases

Definition:

Bases are substances that can accept a proton (H⁺ ion) or donate a hydroxide ion (OH⁻) in aqueous solutions. Bases increase the concentration of hydroxide ions (OH⁻).

Characteristics of Bases:

Taste: Bases generally have a bitter taste and a slippery feel (e.g., soap).

pH: Bases have a pH greater than 7.

Reactivity: Bases react with acids to form salts and water (neutralization reaction).

Types of Bases:

Strong Bases: Fully dissociate into their ions in water, resulting in a high concentration of hydroxide ions. Examples include sodium hydroxide (NaOH) and potassium hydroxide (KOH).

Weak Bases: Partially dissociate in water, resulting in a lower concentration of hydroxide ions. Examples include ammonia (NH₃) and magnesium hydroxide (Mg(OH)₂).

Examples in Biology:

Bicarbonate Ion (HCO₃⁻): Acts as a buffer in blood to help maintain pH balance.

Ammonia: Produced as a waste product in the metabolism of proteins and is neutralized in the body.

3. pH Scale

Definition:

The pH scale measures the concentration of hydrogen ions (H⁺) in a solution, indicating its acidity or basicity. The scale ranges from 0 to 14.

Scale Details:

Acidic Solutions: Have a pH less than 7. The lower the pH, the higher the concentration of hydrogen ions.

Neutral Solutions: Have a pH of 7. Pure water is an example, where the concentration of hydrogen ions equals the concentration of hydroxide ions.

Basic Solutions: Have a pH greater than 7. The higher the pH, the lower the concentration of hydrogen ions.

pH Calculation:

pH Formula: $\text{pH} = -\log[\text{H}^+]$, where [H⁺] is the concentration of hydrogen ions in moles per liter.

pOH Formula: $\text{pOH} = -\log[\text{OH}^-]$, where [OH⁻] is the concentration of hydroxide ions. The relationship between pH and pOH is $\text{pH} + \text{pOH} = 14$.

Importance in Biology:

Enzyme Function: Enzymes have optimal pH ranges in which they function best. Deviations from this range can reduce enzyme activity or denature the enzyme.

Cellular Processes: Many biochemical processes, including metabolic reactions and the transport of substances across cell membranes, are pH-dependent.

4. Buffer Systems

Definition:

Buffers are solutions that resist changes in pH when small amounts of acid or base are added. They help maintain a relatively constant pH in biological systems.

Components:

Weak Acid and Its Conjugate Base: The weak acid can neutralize added bases, while its conjugate base can neutralize added acids. For example, the acetic acid (CH_3COOH) and acetate ion (CH_3COO^-) buffer system.

Buffer Capacity: Refers to the amount of acid or base a buffer solution can neutralize before the pH changes significantly.

Examples in Biology:

Blood Buffer System: The bicarbonate buffer system (H_2CO_3 and HCO_3^-) helps maintain the blood pH around 7.4.

Phosphate Buffer System: Found in intracellular fluids and helps maintain pH stability within cells.

5. Biological Significance

a. Homeostasis

Definition:

Homeostasis is the maintenance of a stable internal environment within an organism, including the regulation of pH.

Importance:

pH Balance: Proper pH levels are crucial for enzyme activity, metabolic processes, and cellular functions. Deviations from normal pH can lead to metabolic disorders or affect the function of cells and tissues.

b. Cellular Respiration and Metabolism

Definition:

Cellular respiration and metabolism involve a series of biochemical reactions that produce energy and require precise pH conditions for optimal enzyme activity.

Importance:

Energy Production: Metabolic processes such as glycolysis and the citric acid cycle are sensitive to pH changes. Maintaining appropriate pH levels ensures efficient energy production and prevents metabolic disturbances.

c. Environmental Impact

Definition:

Environmental changes can affect the pH of ecosystems, influencing the health and survival of organisms.

Importance:

Acid Rain: Acid rain, caused by the release of sulfur dioxide (SO_2) and nitrogen oxides (NO_x) into the atmosphere, can lower the pH of soil and water, harming aquatic life and vegetation.

Ocean Acidification: The absorption of excess atmospheric CO_2 by oceans lowers the pH of seawater, affecting marine organisms, particularly those with calcium carbonate shells or skeletons.

Understanding acids, bases, and pH is crucial for comprehending the chemical basis of biological systems. The properties of acids and bases, the pH scale, and buffer systems are essential for maintaining homeostasis, supporting enzymatic functions, and ensuring proper metabolic processes. These concepts are fundamental to both basic biology and applied fields such as medicine, environmental science, and biochemistry.

Part II: The Building Blocks of Life

Chapter 4: Cell Structure and Function

The Cell Theory

The **Cell Theory** is one of the foundational principles of biology. It provides a framework for understanding the structure and function of living organisms at the most basic level. The cell theory emerged through centuries of scientific discovery and refinement, and it remains a fundamental concept in modern biology.

1. Historical Development of the Cell Theory

The development of the cell theory is a story of scientific observation, experimentation, and refinement over centuries. Here are some key milestones in its development:

a. Discovery of Cells:

Robert Hooke (1665): Robert Hooke was the first scientist to observe and describe cells. Using a primitive microscope, he examined a thin slice of cork and noticed small, box-like structures, which he called "cells" because they reminded him of the small rooms, or cells, in a monastery. However, Hooke was observing the cell walls of dead plant tissue, not living cells.

Anton van Leeuwenhoek (1674): A few years after Hooke's discovery, Anton van Leeuwenhoek, a Dutch scientist, used a more refined microscope to observe living cells for the first time. He described various single-celled organisms, which he called "animalcules," including protozoa and bacteria, in water samples. He also observed human red blood cells.

b. Development of Cell Theory:

Matthias Schleiden (1838): A German botanist, Schleiden, concluded that all plant tissues are composed of cells. He proposed that cells are the fundamental unit of life in plants, suggesting that all plant growth and development occur through cell division.

Theodor Schwann (1839): Shortly after Schleiden's findings, Theodor Schwann, a German zoologist, made a similar conclusion regarding animal tissues. Schwann extended Schleiden's theory to animals, suggesting that all living things, whether plants or animals, are made up of cells. This idea formed the basis for the generalization that cells are the building blocks of all living organisms.

Rudolf Virchow (1855): Virchow, a German physician, made a significant contribution to the cell theory by proposing that all cells arise from pre-existing cells. His famous statement, *Omnis cellula e cellula* (Latin for "All cells come from cells"), challenged the prevailing belief in spontaneous generation (the idea that life could arise from non-living matter) and helped solidify the third tenet of cell theory.

2. The Three Main Tenets of the Cell Theory

The cell theory, as it stands today, is based on three core principles:

All living organisms are composed of one or more cells.

This principle establishes that the cell is the fundamental unit of life. Whether an organism is unicellular (consisting of a single cell, like bacteria) or multicellular (composed of many cells, like humans), all living beings are made up of cells.

The cell is the basic unit of structure and function in all living organisms.

Cells are the smallest functional units of life. They carry out the processes necessary for life, such as metabolism, growth, and reproduction. In multicellular organisms, cells form tissues, which then make up organs and organ systems.

All cells arise from pre-existing cells.

New cells are produced by the division of existing cells. This process of cell division is crucial for growth, development, and reproduction in organisms. It refutes the idea of spontaneous generation and supports the continuity of life.

3. Modern Additions to the Cell Theory

As scientific knowledge expanded, the basic cell theory was further refined with additional principles that address the molecular and functional aspects of cells:

Cells contain hereditary information (DNA) that is passed on during cell division.

This idea recognizes the role of DNA as the genetic material within cells, which is replicated and distributed to daughter cells during cell division, ensuring that the offspring inherit the same genetic information as the parent cells.

All cells are basically the same in chemical composition and metabolic activities.

Although cells can vary greatly in form and function, the fundamental chemical and metabolic processes, such as respiration, energy production, and biosynthesis, are similar across all cells.

Energy flow (metabolism and biochemistry) occurs within cells.

Cells are the sites of all biochemical processes, including the generation and use of energy through metabolic pathways like cellular respiration and photosynthesis. These processes enable cells to perform their functions and sustain life.

4. Importance of the Cell Theory

The cell theory is central to biology because it provides a framework for understanding how all living organisms are organized and how life functions at the most basic level. Its significance lies in several key areas:

a. Unifying Concept in Biology:

The cell theory unifies biology by providing a common denominator for all living organisms. Whether studying microbes, plants, or animals, biologists recognize that cells are the fundamental unit of life. This commonality allows for the study of life across diverse organisms.

b. Basis for Understanding Disease:

Understanding that all cells arise from pre-existing cells is crucial for fields like medicine and pathology. Diseases such as cancer are a result of abnormal cell growth and division. By studying cells, scientists can learn about the underlying causes of diseases, leading to the development of treatments and therapies.

c. Basis for Biotechnology:

The cell theory underpins modern biotechnology, including genetic engineering, stem cell research, and cloning. Understanding the structure and function of cells allows scientists to manipulate genetic material, develop new medicines, and explore therapeutic interventions for various diseases.

d. Cellular Reproduction and Growth:

The cell theory emphasizes that all new cells arise from pre-existing cells. This understanding is fundamental to the study of growth and reproduction, both in individual organisms and in populations. The cell cycle, including processes like mitosis and meiosis, explains how organisms grow and reproduce.

5. Types of Cells: Prokaryotic vs. Eukaryotic

The cell theory applies to all forms of life, but there are two broad categories of cells: prokaryotic and eukaryotic cells. Understanding the differences between these types of cells is key to comprehending the diversity of life on Earth.

a. Prokaryotic Cells:

Definition: Prokaryotic cells are simpler, smaller cells without a true nucleus or membrane-bound organelles. The genetic material (DNA) is not enclosed within a nuclear membrane.

Examples: Bacteria and Archaea.

Characteristics:

Lack a membrane-bound nucleus.

Contain ribosomes for protein synthesis.

May have cell walls, flagella, or pili.

Reproduce through binary fission.

b. Eukaryotic Cells:

Definition: Eukaryotic cells are more complex and contain a true nucleus, which houses the cell's genetic material, as well as membrane-bound organelles.

Examples: Plants, animals, fungi, and protists.

Characteristics:

Have a defined nucleus containing the cell's DNA.

Contain organelles such as the mitochondria, endoplasmic reticulum, Golgi apparatus, and in plants, chloroplasts.

Typically larger and more complex than prokaryotic cells.

Reproduce through mitosis (in somatic cells) and meiosis (in germ cells).

6. Applications of the Cell Theory in Modern Science

The cell theory has profound implications and applications in modern science. Here are some of the most significant areas where the cell theory plays a critical role:

a. Medicine:

Stem Cell Research: Understanding that all cells arise from pre-existing cells has opened the door for stem cell research, where unspecialized cells can be induced to become specific cell types for therapies and regeneration.

Cancer Research: Cancer is fundamentally a disease of uncontrolled cell growth. By studying how normal cells divide and how cancer cells differ, scientists can develop treatments that target these abnormalities.

b. Genetic Engineering:

The cell theory forms the basis of genetic engineering, where scientists modify the genetic material inside cells to produce new traits, create transgenic organisms, or produce therapeutic proteins.

c. Evolutionary Biology:

The cell theory supports evolutionary theory by showing that all living organisms share a common cellular ancestry. Studies of prokaryotic and eukaryotic cells offer insights into the evolution of complex life forms.

The cell theory remains one of the most important concepts in biology. By stating that all living organisms are made of cells, cells are the basic unit of life, and all cells come from pre-existing cells, the theory provides a unifying principle that underpins the study of life. It forms the foundation for modern biological sciences, from genetics and medicine to evolution and biotechnology. Understanding the cell theory is essential for comprehending the structure and function of living organisms and the complex processes that sustain life.

Prokaryotic vs. Eukaryotic Cells

Understanding the differences between prokaryotic and eukaryotic cells is fundamental to grasping the diversity and complexity of life. Both types of cells are the basic structural and functional units of organisms, but they differ significantly in their organization and complexity. This chapter explores these differences extensively.

1. Overview of Prokaryotic Cells

Definition:

Prokaryotic cells are simpler, smaller cells without a true nucleus or membrane-bound organelles. They represent the most ancient form of cellular life and are found in two major domains: Bacteria and Archaea.

Characteristics:

a. Genetic Material:

Nucleoid Region: Prokaryotic cells lack a membrane-bound nucleus. Instead, their genetic material (DNA) is located in a region called the nucleoid, which is not enclosed by a membrane.

Chromosomes: Prokaryotes typically have a single, circular chromosome that contains the essential genetic information. They may also have smaller, circular DNA molecules called plasmids that carry additional genetic information.

b. Cellular Structure:

Cell Wall: Most prokaryotes have a rigid cell wall that provides structural support and protection. In bacteria, the cell wall contains peptidoglycan, while in archaea, it may be composed of different substances.

Plasma Membrane: Beneath the cell wall, prokaryotes have a plasma membrane that controls the movement of substances in and out of the cell.

Ribosomes: Prokaryotic cells contain ribosomes for protein synthesis. These are smaller than eukaryotic ribosomes but perform the same function.

c. Organelles:

Lack of Membrane-Bound Organelles: Prokaryotes do not have membrane-bound organelles such as mitochondria, chloroplasts, or the endoplasmic reticulum. Their metabolic processes occur within the cytoplasm or at the plasma membrane.

Cytoskeleton: Prokaryotic cells have a simpler cytoskeleton compared to eukaryotes, which helps maintain cell shape and facilitate movement.

d. Reproduction:

Binary Fission: Prokaryotic cells reproduce asexually through a process called binary fission, where the cell divides into two identical daughter cells.

e. Examples:

Bacteria: E. coli, Salmonella, and Staphylococcus.

Archaea: Methanogens, Halophiles, and Thermophiles.

2. Overview of Eukaryotic Cells

Definition:

Eukaryotic cells are more complex and larger than prokaryotic cells. They possess a true nucleus and various membrane-bound organelles. Eukaryotes include plants, animals, fungi, and protists.

Characteristics:

a. Genetic Material:

Nucleus: Eukaryotic cells have a well-defined nucleus enclosed by a nuclear envelope, which contains the cell's genetic material (DNA). The nucleus is the control center of the cell.

Chromosomes: Eukaryotes have multiple, linear chromosomes organized within the nucleus. DNA is wrapped around histone proteins to form chromatin.

b. Cellular Structure:

Cell Membrane: Eukaryotic cells have a plasma membrane that regulates the movement of substances in and out of the cell.

Cell Wall: Present in plant cells and fungi but not in animal cells. Plant cell walls are made of cellulose, while fungal cell walls are made of chitin.

c. Organelles:

Nucleus: Contains the genetic material and is the site of DNA replication and transcription.

Mitochondria: Known as the powerhouse of the cell, mitochondria are involved in ATP production through cellular respiration.

Chloroplasts: Found in plant cells and some protists, chloroplasts are responsible for photosynthesis and contain chlorophyll.

Endoplasmic Reticulum (ER): The ER is involved in protein and lipid synthesis. It comes in two forms: rough ER (with ribosomes) and smooth ER (without ribosomes).

Golgi Apparatus: Modifies, sorts, and packages proteins and lipids for secretion or delivery to other organelles.

Lysosomes and Peroxisomes: Lysosomes contain digestive enzymes to break down waste materials, while peroxisomes are involved in oxidative reactions and detoxification.

Vacuoles: Storage organelles that can hold nutrients, waste products, or help maintain turgor pressure in plant cells.

d. Reproduction:

Mitosis and Meiosis: Eukaryotic cells reproduce through mitosis (for growth and repair) and meiosis (for the production of gametes in sexual reproduction).

e. Examples:

Animals: Human cells, cells of mammals, birds, etc.

Plants: Cells of trees, flowers, and grasses.

Fungi: Yeast, mushrooms.

Protists: Amoebas, algae.

3. Comparative Analysis

a. Size and Complexity:

Prokaryotic Cells: Generally smaller (0.1-5 µm in diameter) and simpler in structure.

Eukaryotic Cells: Larger (10-100 µm in diameter) and more complex, with a greater variety of organelles.

b. Genetic Material Organization:

Prokaryotic Cells: DNA is circular and located in the nucleoid region without a membrane.

Eukaryotic Cells: DNA is linear and enclosed within a membrane-bound nucleus.

c. Organelles and Structures:

Prokaryotic Cells: Lack membrane-bound organelles, have a simpler cytoskeleton.

Eukaryotic Cells: Have membrane-bound organelles and a more complex cytoskeleton.

d. Reproduction:

Prokaryotic Cells: Reproduce by binary fission, which is a simple and rapid process.

Eukaryotic Cells: Reproduce by mitosis (asexual) or meiosis (sexual), involving more complex processes and regulation.

4. Evolutionary Perspectives

a. Origin:

Prokaryotic Cells: Considered the earliest form of life on Earth. The theory suggests that prokaryotic cells evolved first and that eukaryotic cells evolved later through endosymbiosis.

Eukaryotic Cells: Believed to have evolved from prokaryotic ancestors through a process known as endosymbiotic theory. This theory proposes that eukaryotic cells originated from a symbiotic relationship between early eukaryotes and engulfed prokaryotic cells, which eventually became organelles like mitochondria and chloroplasts.

b. Endosymbiotic Theory:

Mitochondria and Chloroplasts: These organelles have their own DNA and double membranes, supporting the idea that they originated from free-living prokaryotes that were engulfed by ancestral eukaryotic cells.

5. Applications in Science and Medicine

a. Antibiotics and Treatment:

Prokaryotic Cells: Antibiotics target specific features of bacterial cells, such as the cell wall or ribosomes, which are absent in eukaryotic cells. This specificity helps in treating bacterial infections without harming human cells.

b. Genetic Engineering:

Eukaryotic Cells: Understanding eukaryotic cell biology is crucial for genetic engineering, including gene therapy and recombinant DNA technology. For example, genetically modified plants and animals are produced using techniques that rely on knowledge of eukaryotic cell structure and function.

c. Evolutionary Studies:

Comparative studies of prokaryotic and eukaryotic cells provide insights into the evolution of life and the relationships between different forms of life. These studies help in understanding the origin of complex cellular processes and the evolutionary history of organisms.

The comparison between prokaryotic and eukaryotic cells highlights the diversity and complexity of life at the cellular level. Prokaryotic cells, with their simpler structure and organization, represent the earliest forms of life, while eukaryotic cells, with their complex organelles and compartmentalization, illustrate the advanced level of cellular organization found in plants, animals, fungi, and protists. Understanding these differences is essential for fields ranging from microbiology to biotechnology and evolutionary biology.

Cell Organelles and Their Functions

Cells are the basic units of life, and within each cell, there are specialized structures known as **organelles** that perform distinct functions essential to the survival, growth, and reproduction of the cell. Organelles are responsible for maintaining cellular homeostasis, producing energy, synthesizing proteins, and managing waste, among other tasks.

Organelles can be broadly categorized into **membrane-bound** (mostly found in eukaryotic cells) and **non-membrane-bound** (found in both eukaryotic and prokaryotic cells). Below is an extensive discussion of the various cell organelles and their functions.

1. Nucleus

Function: The nucleus is the control center of the cell, housing the cell's genetic material (DNA). It directs all cellular activities, including growth, metabolism, protein synthesis, and reproduction.

Structure: The nucleus is surrounded by a double membrane called the **nuclear envelope**, which contains nuclear pores for the passage of molecules like RNA and proteins.

Nucleolus: Inside the nucleus, the nucleolus is the site of **ribosome synthesis**. It assembles ribosomal RNA (rRNA) and proteins to form ribosomes, which are essential for protein production.

Chromatin: The DNA in the nucleus is organized into chromatin, a complex of DNA and proteins, which condenses into chromosomes during cell division.

2. Ribosomes

Function: Ribosomes are the molecular machines that synthesize proteins by translating messenger RNA (mRNA) into polypeptide chains. They are essential for the production of proteins required for cellular functions.

Structure: Ribosomes consist of two subunits, a large and a small subunit, each made of rRNA and proteins.

Location: Ribosomes can either be **free-floating** in the cytoplasm or attached to the rough endoplasmic reticulum (ER). Free ribosomes typically synthesize proteins that function within the cytoplasm, while those on the rough ER synthesize proteins destined for secretion or for use in membranes and organelles.

3. Endoplasmic Reticulum (ER)

The endoplasmic reticulum is an extensive network of membranes involved in protein and lipid synthesis and transport.

a. Rough Endoplasmic Reticulum (Rough ER)

Function: The rough ER is studded with ribosomes, giving it a rough appearance. It is primarily involved in the **synthesis of proteins**, particularly those that are destined for secretion, incorporation into the plasma membrane, or use within other organelles like lysosomes.

Protein Processing: Proteins synthesized on the rough ER are folded and modified within its lumen before being transported to the Golgi apparatus.

b. Smooth Endoplasmic Reticulum (Smooth ER)

Function: The smooth ER lacks ribosomes and is involved in the **synthesis of lipids** (including phospholipids and steroids), **detoxification of harmful substances**, and **storage of calcium ions** (especially in muscle cells).

Carbohydrate Metabolism: The smooth ER is also involved in carbohydrate metabolism, including the conversion of glucose-6-phosphate to glucose.

4. Golgi Apparatus (Golgi Complex)

Function: The Golgi apparatus is the cell's "post office." It modifies, sorts, and packages proteins and lipids received from the ER for transport to their final destinations. It is involved in the **addition of carbohydrates (glycosylation)** to proteins, forming glycoproteins.

Structure: The Golgi consists of flattened membranous sacs called **cisternae**, and it has a receiving face (cis-face) and a shipping face (trans-face).

Vesicles: Packaged proteins and lipids are transported from the Golgi in vesicles to their target locations, such as the plasma membrane, lysosomes, or extracellular space.

5. Lysosomes

Function: Lysosomes are the cell's digestive system. They contain **hydrolytic enzymes** that break down macromolecules (proteins, lipids, carbohydrates, and nucleic acids), damaged organelles, and foreign particles such as bacteria. This process is called **autophagy**.

Structure: Lysosomes are membrane-bound vesicles filled with enzymes that function optimally at acidic pH levels.

6. Peroxisomes

Function: Peroxisomes are small, membrane-bound organelles that contain enzymes responsible for breaking down long-chain fatty acids through **β-oxidation**, and detoxifying harmful substances like hydrogen peroxide (H_2O_2) into water and oxygen using the enzyme **catalase**.

Metabolic Role: Peroxisomes also play a role in the metabolism of amino acids and the synthesis of cholesterol and bile acids.

7. Mitochondria

Function: Mitochondria are the **powerhouses of the cell**, generating energy in the form of **adenosine triphosphate (ATP)** through the process of **cellular respiration**. They play a crucial role in energy metabolism, the regulation of the cell cycle, and programmed cell death (apoptosis).

Structure: Mitochondria have a double membrane. The **outer membrane** is smooth, while the **inner membrane** is highly folded into structures called **cristae**, which increase surface area for ATP production. The inner membrane encloses the **mitochondrial matrix**, which contains enzymes, DNA, and ribosomes.

Endosymbiotic Theory: Mitochondria are thought to have originated from free-living prokaryotes that entered into a symbiotic relationship with ancestral eukaryotic cells.

8. Chloroplasts (in Plant Cells)

Function: Chloroplasts are the site of **photosynthesis** in plant cells, converting light energy into chemical energy stored in the form of glucose. They contain the green pigment **chlorophyll**, which captures light energy.

Structure: Chloroplasts have a double membrane and contain disc-like structures called **thylakoids**, stacked into **grana**. The thylakoid membranes contain chlorophyll, and the surrounding fluid is known as the **stroma**, where the Calvin cycle (carbon fixation) occurs.

Endosymbiotic Theory: Like mitochondria, chloroplasts are believed to have originated from free-living prokaryotes (cyanobacteria) through endosymbiosis.

9. Vacuoles

Function: Vacuoles are storage organelles found in both plant and animal cells, though they are more prominent in plant cells. They store nutrients, waste products, and help maintain **turgor pressure** in plant cells, which is important for structural support.

Structure: In plant cells, the central vacuole can occupy up to 90% of the cell's volume, storing water, ions, and other molecules.

Contractile Vacuole: In some protists, contractile vacuoles help regulate water balance by expelling excess water from the cell.

10. Cytoskeleton

Function: The cytoskeleton provides structural support, shape, and movement for the cell. It is composed of three main types of protein filaments:

Microfilaments (Actin Filaments): Involved in cell movement, cytokinesis (cell division), and muscle contraction.

Intermediate Filaments: Provide mechanical strength and help maintain the shape of the cell.

Microtubules: Form the mitotic spindle during cell division, provide tracks for intracellular transport, and are involved in the structure of cilia and flagella.

Cytoskeletal Dynamics: The cytoskeleton is highly dynamic, constantly reorganizing to accommodate the cell's needs.

11. Centrosomes and Centrioles

Function: Centrosomes serve as the main microtubule organizing centers (MTOCs) in animal cells. They play a crucial role during **cell division**, organizing the mitotic spindle that separates chromosomes.

Centrioles: Located within the centrosome, centrioles are cylindrical structures made of microtubules. They are involved in the formation of **cilia** and **flagella**, as well as organizing the microtubules during mitosis.

12. Plasma Membrane (Cell Membrane)

Function: The plasma membrane is a selective barrier that regulates the movement of substances in and out of the cell. It maintains **homeostasis** by controlling what enters and exits the cell.

Structure: The plasma membrane is composed of a **phospholipid bilayer** with embedded proteins, cholesterol, and carbohydrates. This structure is often described by the **fluid mosaic model**, where the components are dynamic and fluid-like.

Transport Proteins: Help in the movement of molecules across the membrane via **passive transport** (diffusion, facilitated diffusion) and **active transport** (pumps and vesicular transport).

Receptors: Proteins that receive signals from outside the cell and initiate cellular responses.

13. Cilia and Flagella

Function: Cilia and flagella are hair-like structures that extend from the cell surface and are involved in movement.

Cilia: Short and numerous, cilia are responsible for moving substances across the surface of the cell or aiding in the locomotion of single-celled organisms like **paramecium**. In human respiratory tracts, cilia help sweep mucus and trapped debris out of the lungs.

Flagella: Longer and usually fewer in number, flagella are used for cell movement. For example, the **sperm cell** moves using a single flagellum. Both cilia and flagella are composed of **microtubules** arranged in a "9+2" pattern, with nine pairs of microtubules surrounding two central microtubules.

14. Cell Wall (in Plant Cells, Fungi, and Some Prokaryotes)

Function: The cell wall provides structural support and protection to cells. It also helps maintain the cell's shape and prevents excessive water intake through **osmosis**.

Structure:

Plants: The plant cell wall is primarily made of **cellulose**, a strong carbohydrate polymer that gives rigidity to plant cells.

Fungi: The cell wall in fungi is made of **chitin**, another strong structural polysaccharide.

Prokaryotes (Bacteria): The bacterial cell wall contains **peptidoglycan**, a polymer that provides structural integrity and protects against changes in osmotic pressure.

15. Extracellular Matrix (ECM)

Function: The extracellular matrix is a complex network of proteins and carbohydrates surrounding animal cells. It provides structural and biochemical support to surrounding cells, facilitates communication between cells, and helps cells stick together to form tissues.

Components: Major components of the ECM include **collagen** (for strength), **elastin** (for elasticity), **proteoglycans**, and **fibronectin**. Cells interact with the ECM through transmembrane proteins called **integrins**.

Organelle	Function
Nucleus	Houses DNA; controls cell activities; site of transcription
Ribosomes	Protein synthesis
Rough ER	Protein folding and modification; transport to Golgi
Smooth ER	Lipid synthesis; detoxification; calcium storage
Golgi Apparatus	Modifies, sorts, and packages proteins and lipids
Lysosomes	Digestion of macromolecules; breakdown of worn-out organelles
Peroxisomes	Breakdown of fatty acids; detoxification of harmful compounds (e.g., hydrogen peroxide)
Mitochondria	ATP production via cellular respiration; energy generation
Chloroplasts	Photosynthesis in plant cells
Vacuoles	Storage of substances; maintain turgor pressure in plant cells
Cytoskeleton	Structural support; intracellular transport; cell division; movement
Centrosomes and Centrioles	Organize microtubules during cell division
Plasma Membrane	Regulates entry and exit of substances; maintains cell integrity
Cilia and Flagella	Movement of cells or substances across cell surfaces
Cell Wall	Structural support and protection in plant, fungi, and some bacterial cells
Extracellular Matrix	Structural support; communication and adhesion

Cell organelles are specialized structures that perform vital functions, making life at the cellular level possible. Understanding the role of each organelle is critical to the study of biology as they collectively contribute to the growth, survival, reproduction, and overall functioning of cells. While many organelles are common to all eukaryotic cells, others, like chloroplasts and cell walls, are unique to plant cells. The distinction between these organelles and their roles highlights the complexity and efficiency of cellular machinery in sustaining life.

Cell Membrane Structure and Function

The **cell membrane**, also known as the **plasma membrane**, is a crucial component of both prokaryotic and eukaryotic cells. It acts as a dynamic barrier between the interior of the cell and its external environment, maintaining the structural integrity of the cell while regulating the movement of substances in and out of the cell. This selective permeability ensures the cell can maintain homeostasis—an optimal internal environment required for its survival.

In this discussion, we will cover the structure of the cell membrane, its various functions, and the importance of its dynamic nature in cellular processes.

1. Structure of the Cell Membrane

The **Fluid Mosaic Model** is the most widely accepted model that explains the structure of the cell membrane. It describes the membrane as a fluid and dynamic structure made up of several components, with proteins and other molecules embedded within a **phospholipid bilayer**. The components are constantly moving, allowing flexibility and interaction.

a. Phospholipid Bilayer

Phospholipids are the primary structural molecules of the membrane. Each phospholipid has a **hydrophilic (water-attracting) head** and two **hydrophobic (water-repelling) tails**.

When exposed to an aqueous environment, phospholipids arrange themselves into a bilayer. The **hydrophilic heads** face outward, interacting with the watery environments both inside (cytoplasm) and outside the cell, while the **hydrophobic tails** face inward, away from water, forming the core of the membrane.

This bilayer forms the basic structural framework of the membrane, creating a semipermeable barrier that regulates the movement of substances based on size, charge, and solubility.

b. Membrane Proteins

Proteins embedded in or associated with the phospholipid bilayer serve numerous functions, making the membrane more than just a passive barrier.

Integral Proteins: These proteins are embedded within the lipid bilayer, often spanning the entire membrane. They serve as channels, carriers, and receptors.

Channel Proteins: Form pores in the membrane that allow specific ions or molecules to pass through via **facilitated diffusion**. For example, sodium (Na+) and potassium (K+) ions move across the membrane through these channels.

Carrier Proteins: These bind to specific molecules and undergo a change in shape to transport the molecules across the membrane. This can happen through either **facilitated diffusion** or **active transport**.

Receptor Proteins: Serve as binding sites for signaling molecules (ligands) such as hormones, initiating a cellular response. This is crucial for processes like signal transduction and cellular communication.

Peripheral Proteins: These are loosely attached to the surface of the membrane, often interacting with integral proteins or the lipid bilayer. They function in signal transduction, maintaining the cell's shape, and cellular recognition.

c. Carbohydrates

Carbohydrates are often attached to proteins (forming **glycoproteins**) or lipids (forming **glycolipids**) on the extracellular surface of the membrane. These carbohydrates play a significant role in:

Cell recognition: They act as identification markers that allow cells to recognize each other, which is crucial in immune responses.

Adhesion: Carbohydrates can help cells stick together, aiding in tissue formation.

d. Cholesterol

Cholesterol molecules are interspersed within the phospholipid bilayer, especially in animal cells. They provide **stability** to the membrane while maintaining its **fluidity**. Cholesterol prevents the membrane from becoming too rigid in cold temperatures or too fluid in warm temperatures, ensuring that the membrane functions effectively under varying environmental conditions.

2. Functions of the Cell Membrane

The cell membrane plays a variety of critical roles in maintaining the overall function and health of the cell. Its structure allows it to control the movement of substances, mediate communication between cells, and facilitate various biochemical reactions.

a. Selective Permeability

The plasma membrane is **selectively permeable**, meaning it allows certain molecules to pass while blocking others. This selectivity is crucial for maintaining the internal conditions of the cell (homeostasis). The permeability of the membrane depends on:

Molecule Size: Small, non-polar molecules like oxygen (O_2) and carbon dioxide (CO_2) can easily pass through the lipid bilayer via simple diffusion. Larger molecules like glucose or charged ions require transport proteins.

Solubility: Nonpolar, hydrophobic molecules can diffuse through the lipid bilayer, while polar, hydrophilic molecules usually need help from transport proteins.

Charge: Ions such as Na^+ and K^+ cannot freely cross the membrane due to their charge; they require specialized ion channels or pumps for movement.

b. Transport Across the Membrane

The cell membrane regulates the movement of substances using **passive** and **active transport** mechanisms.

Passive Transport: Movement of molecules across the membrane without the use of cellular energy (ATP).

Diffusion: The movement of molecules from an area of high concentration to an area of low concentration. Oxygen and carbon dioxide diffuse directly across the membrane.

Facilitated Diffusion: Uses transport proteins (like channel or carrier proteins) to move larger or polar molecules (e.g., glucose) down their concentration gradient without energy input.

Osmosis: A specific type of facilitated diffusion, osmosis is the movement of water across a selectively permeable membrane via aquaporins (water channels).

Active Transport: The movement of molecules against their concentration gradient, requiring energy (ATP).

Sodium-Potassium Pump (Na⁺/K⁺ Pump): This pump actively transports Na⁺ out of the cell and K⁺ into the cell, creating an electrochemical gradient essential for nerve impulse transmission and muscle contraction.

Endocytosis and Exocytosis: Large molecules (e.g., proteins, polysaccharides) or entire cells (e.g., bacteria) are transported via vesicular transport.

Endocytosis: The cell engulfs materials by wrapping its membrane around them and forming a vesicle. There are two types:

Phagocytosis: "Cell eating," where large particles are engulfed.

Pinocytosis: "Cell drinking," where the cell takes in fluids and dissolved substances.

Exocytosis: The process by which vesicles fuse with the membrane to release substances outside the cell. This is crucial for secretion of hormones and neurotransmitters.

c. Cell Communication

The plasma membrane plays a vital role in communication between cells through **cell signaling** and **receptor-mediated signaling**.

Signal Transduction: Membrane receptors detect chemical signals (e.g., hormones or neurotransmitters) from the external environment. When a signaling molecule binds to a receptor, it triggers a cascade of intracellular events that lead to a cellular response.

Receptor-Mediated Endocytosis: Cells absorb specific molecules by engulfing them with the help of membrane receptors. For example, **LDL (low-density lipoprotein)** particles are taken up by cells through receptor-mediated endocytosis.

d. Cell Adhesion and Recognition

Cell Adhesion: The membrane plays an essential role in the attachment of cells to each other, which is crucial for the formation of tissues and organs.

Tight Junctions: Create a watertight seal between cells.

Desmosomes: Anchor cells to one another, providing mechanical stability.

Gap Junctions: Create channels between adjacent cells, allowing the passage of small molecules and ions.

Cell Recognition: Carbohydrates on the cell surface act as molecular signatures that cells use to recognize each other, facilitating tissue formation, immune responses, and organ development.

e. Structural Support

While the **cytoskeleton** provides the internal framework of the cell, the plasma membrane helps in organizing the cytoskeleton and facilitating its connection with the extracellular matrix (ECM). This support is crucial for maintaining the cell's shape, especially in cells that are subject to mechanical stress, such as muscle and skin cells.

f. Electrochemical Gradients

The cell membrane helps establish and maintain **electrochemical gradients** across its surface. These gradients are critical for numerous cellular processes, including:

Nerve Signal Transmission: The Na$^+$/K$^+$ pump maintains a voltage difference (membrane potential) across nerve cell membranes, which is essential for the propagation of electrical signals.

Muscle Contraction: The movement of calcium ions (Ca^{2+}) across membranes triggers muscle contraction.

3. Importance of Membrane Fluidity

Membrane **fluidity** is essential for the function of the cell membrane. Factors that influence membrane fluidity include:

Cholesterol: Cholesterol molecules inserted between phospholipids help to stabilize the membrane by preventing the fatty acid tails from sticking together, ensuring that the membrane remains fluid and flexible.

Fatty Acid Composition: The presence of unsaturated fatty acids in the phospholipids increases fluidity, while saturated fatty acids make the membrane more rigid.

Temperature: Lower temperatures make membranes more rigid, while higher temperatures increase fluidity. Cells must regulate their membrane composition to adapt to changes in temperature.

The fluidity of the membrane is critical for its functions, including:

Lateral movement of proteins: Enables membrane proteins to move around within the membrane, allowing interactions necessary for cell signaling and transport.

Endocytosis and exocytosis: The ability of the membrane to deform and merge with vesicles is dependent on its fluid nature.

Self-healing: The membrane's ability to repair small tears and maintain integrity is aided by its fluidity.

The cell membrane is a complex and dynamic structure crucial for maintaining cellular function and homeostasis. Its composition—phospholipid bilayer, membrane proteins, carbohydrates, and cholesterol—enables it to perform vital roles such as selective permeability, transport, cell communication, adhesion, and structural support. The fluid mosaic model aptly describes the membrane's flexible and ever-changing nature, which is essential for its ability to adapt to varying conditions and meet the cell's needs. Understanding the structure and function of the cell membrane is fundamental to comprehending how cells interact with their environment and maintain their internal equilibrium.

Chapter 6: Cellular Processes
Diffusion and Osmosis

Diffusion and **osmosis** are fundamental cellular processes that involve the movement of molecules across cell membranes. Both processes are essential for maintaining cellular homeostasis and facilitating various physiological functions. Here, we will discuss these processes extensively, covering their mechanisms, types, factors influencing their rates, and their significance in biological systems.

1. Diffusion

Diffusion is the movement of molecules or ions from an area of higher concentration to an area of lower concentration until equilibrium is achieved. This process occurs due to the natural tendency of molecules to spread out and fill available space.

a. Mechanism of Diffusion

Concentration Gradient: Molecules move along their concentration gradient, which is the difference in concentration of a substance between two regions. Diffusion continues until the concentration is equal on both sides, reaching **dynamic equilibrium**.

Passive Process: Diffusion does not require energy input (ATP) because it relies on the kinetic energy of molecules. It is a type of **passive transport**.

b. Types of Diffusion

Simple Diffusion: Involves the movement of small, nonpolar molecules (e.g., oxygen, carbon dioxide) directly through the phospholipid bilayer of the cell membrane. These molecules can diffuse across the membrane because they are soluble in the lipid bilayer.

Facilitated Diffusion: Involves the movement of larger or polar molecules (e.g., glucose, ions) through specific transport proteins embedded in the membrane. There are two main types:

Channel Proteins: Form pores that allow specific ions or molecules to pass through. For example, potassium channels allow the passage of K^+ ions.

Carrier Proteins: Bind to specific molecules and undergo conformational changes to transport the molecules across the membrane. For example, glucose transporters facilitate the entry of glucose into cells.

c. Factors Affecting Diffusion

Concentration Gradient: The greater the difference in concentration between two regions, the faster the rate of diffusion.

Temperature: Higher temperatures increase molecular movement, thus increasing the rate of diffusion.

Size of Molecules: Smaller molecules diffuse more quickly than larger molecules.

Surface Area: A larger surface area of the membrane allows for more molecules to diffuse simultaneously.

Distance: Shorter distances (i.e., thinner membranes) increase the rate of diffusion.

d. Examples of Diffusion in Biology

Respiratory Gas Exchange: Oxygen diffuses from the alveoli in the lungs (where its concentration is high) into the blood (where its concentration is lower). Conversely, carbon dioxide diffuses from the blood into the alveoli to be exhaled.

Nutrient Absorption: In the intestines, nutrients such as amino acids and sugars diffuse into the blood from the digestive tract.

2. Osmosis

Osmosis is a specific type of diffusion that involves the movement of water molecules through a selectively permeable membrane from an area of lower solute concentration (higher water concentration) to an area of higher solute concentration (lower water concentration).

a. Mechanism of Osmosis

Selective Permeability: The cell membrane is selectively permeable to water due to the presence of **aquaporins**, which are specialized water channels that facilitate the movement of water molecules.

Osmotic Pressure: Osmosis creates osmotic pressure, which is the pressure required to stop the flow of water through the membrane. This pressure can influence cell volume and shape.

b. Types of Solutions

Isotonic Solution: The solute concentration inside the cell is equal to that outside the cell. Water moves in and out of the cell at an equal rate, resulting in no net movement of water. Cells maintain their normal shape.

Hypertonic Solution: The solute concentration outside the cell is higher than inside. Water moves out of the cell, causing the cell to shrink or crenate. This can lead to cell dehydration.

Hypotonic Solution: The solute concentration inside the cell is higher than outside. Water moves into the cell, causing it to swell and potentially burst or lyse. This can occur in animal cells but is less problematic in plant cells due to the cell wall.

c. Factors Affecting Osmosis

Concentration Gradient: The greater the difference in solute concentration between the inside and outside of the cell, the faster the rate of osmosis.

Membrane Permeability: The presence and density of aquaporins influence the rate of water movement.

Temperature: Higher temperatures increase the kinetic energy of water molecules, thereby increasing the rate of osmosis.

d. Examples of Osmosis in Biology

Cellular Homeostasis: Cells regulate their internal environment through osmosis. For example, kidney cells use osmosis to maintain water balance in the body.

Plant Turgor Pressure: In plant cells, osmosis creates turgor pressure that supports the plant's structure and rigidity. Water entering the vacuole increases internal pressure, pushing the plasma membrane against the cell wall.

3. Comparison of Diffusion and Osmosis

Aspect	Diffusion	Osmosis
Definition	Movement of molecules from high to low concentration.	Movement of water from low to high solute concentration.
Type of Molecules	Nonpolar molecules (e.g., O_2, CO_2) and some small polar molecules.	Water molecules only.
Membrane Involvement	May or may not involve a membrane.	Involves a selectively permeable membrane.
Energy Requirement	No energy required (passive transport).	No energy required (passive transport).
Transport Proteins	Not always required (simple diffusion).	Requires specific water channels (aquaporins) for facilitated osmosis.

Clinical and Practical Implications

Medical Solutions: Understanding osmotic principles is crucial in medical treatments like intravenous (IV) fluid administration. For example, isotonic solutions are used to rehydrate patients without causing cell shrinkage or swelling.

Dialysis: In patients with kidney failure, dialysis uses osmotic principles to remove waste products from the blood and balance electrolytes.

Food Preservation: Osmosis is used in food preservation methods such as salting or sugaring, which creates hypertonic environments to inhibit microbial growth.

Diffusion and osmosis are fundamental processes essential for cellular function and homeostasis. Diffusion refers to the movement of molecules from high to low concentration, while osmosis specifically involves the movement of water through a selectively permeable membrane. Both processes are passive and do not require energy input. Factors such as concentration gradients, temperature, and membrane permeability influence their rates. Understanding these processes is crucial for appreciating how cells interact with their environment and maintain their internal conditions.

Cellular Respiration

Cellular respiration is a critical biochemical process through which cells extract energy from nutrients to produce adenosine triphosphate (ATP), the primary energy carrier in cells. This process involves a series of complex reactions that convert glucose and other molecules into ATP, carbon dioxide (CO_2), and water (H_2O). Cellular respiration is essential for the survival of all aerobic organisms and occurs in three main stages: glycolysis, the Krebs cycle (Citric Acid Cycle), and oxidative phosphorylation (which includes the electron transport chain and chemiosmosis).

1. Overview of Cellular Respiration

Cellular respiration can be broadly divided into **aerobic respiration** (which requires oxygen) and **anaerobic respiration** (which does not require oxygen). Aerobic respiration is the most efficient form of cellular respiration and produces the maximum amount of ATP. Anaerobic respiration occurs in environments where oxygen is scarce or absent and produces less ATP while generating byproducts like lactic acid or ethanol.

a. Aerobic Respiration

Aerobic respiration consists of three main stages:

Glycolysis

Krebs Cycle (Citric Acid Cycle)

Oxidative Phosphorylation (including the Electron Transport Chain and Chemiosmosis)

2. Glycolysis

Glycolysis is the first stage of cellular respiration and occurs in the cytoplasm of the cell. It breaks down one molecule of glucose into two molecules of pyruvate. This process can be summarized in the following steps:

a. Overview

Location: Cytoplasm

Starting Molecule: Glucose (6-carbon sugar)

Ending Molecules: 2 Pyruvate (3-carbon molecules), 2 ATP, and 2 NADH

b. Phases of Glycolysis

Energy Investment Phase:

Steps: Two ATP molecules are used to phosphorylate glucose and its derivative, fructose-1,6-bisphosphate.

Purpose: This phase prepares glucose for subsequent cleavage.

Cleavage Phase:

Steps: Fructose-1,6-bisphosphate is split into two three-carbon molecules: glyceraldehyde-3-phosphate (G3P) and dihydroxyacetone phosphate (DHAP). DHAP is then converted into G3P.

Purpose: Produces two molecules of G3P that will proceed to the next phase.

Energy Payoff Phase:

Steps: Each G3P molecule is converted into pyruvate, generating four ATP molecules (net gain of 2 ATP) and two NADH molecules.

Purpose: Produces ATP and NADH for further stages of respiration.

c. Products

ATP: 2 (net gain)

NADH: 2

Pyruvate: 2

3. Krebs Cycle (Citric Acid Cycle)

The Krebs Cycle, also known as the Citric Acid Cycle, takes place in the mitochondrial matrix. It processes each pyruvate (converted to acetyl-CoA) to produce energy carriers and waste products.

a. Overview

Location: Mitochondrial Matrix

Starting Molecule: Acetyl-CoA (derived from pyruvate)

Ending Molecules: 2 CO_2, 3 NADH, 1 $FADH_2$, and 1 ATP (per turn of the cycle)

b. Steps of the Krebs Cycle

Formation of Citrate: Acetyl-CoA combines with oxaloacetate to form citrate (6-carbon molecule).

Isomerization: Citrate is rearranged to isocitrate.

Oxidative Decarboxylation: Isocitrate is converted into alpha-ketoglutarate, releasing CO_2 and producing NADH.

Further Decarboxylation: Alpha-ketoglutarate is converted into succinyl-CoA, producing another CO_2 and NADH.

ATP Formation: Succinyl-CoA is converted into succinate, generating ATP (or GTP) through substrate-level phosphorylation.

Oxidation: Succinate is oxidized to fumarate, producing $FADH_2$.

Hydration: Fumarate is converted into malate.

Regeneration: Malate is oxidized to oxaloacetate, generating another NADH.

c. Products

ATP: 1 (per cycle turn, 2 per glucose molecule as two acetyl-CoA enter)

NADH: 3 (per cycle turn, 6 per glucose molecule)

$FADH_2$: 1 (per cycle turn, 2 per glucose molecule)

CO₂: 2 (per cycle turn, 4 per glucose molecule)

4. Oxidative Phosphorylation

Oxidative phosphorylation is the final stage of aerobic respiration and takes place in the inner mitochondrial membrane. It consists of two major components: the electron transport chain and chemiosmosis.

a. Electron Transport Chain (ETC)

Location: Inner mitochondrial membrane

Function: Transfers electrons through a series of protein complexes (Complex I-IV) and mobile electron carriers (ubiquinone and cytochrome c).

Electron Transfer: NADH and FADH₂ donate electrons to the ETC. Electrons move through complexes, releasing energy that pumps protons (H⁺) into the intermembrane space, creating a proton gradient.

Final Electron Acceptor: Electrons are finally transferred to oxygen (O₂), forming water as a byproduct.

Role: The transfer of electrons through the ETC establishes a proton gradient across the membrane, which drives ATP synthesis.

b. Chemiosmosis

Location: Inner mitochondrial membrane

Function: Utilizes the proton gradient created by the ETC to drive ATP synthesis.

Proton Gradient: The proton gradient (proton-motive force) drives protons back into the mitochondrial matrix through ATP synthase.

ATP Synthesis: As protons flow through ATP synthase, the enzyme synthesizes ATP from ADP and inorganic phosphate (Pi).

c. Products

ATP: Approximately 28-34 (from oxidative phosphorylation per glucose molecule)

Water: 2 (formed when electrons combine with O₂ and protons)

5. Anaerobic Respiration

In the absence of oxygen, cells rely on anaerobic respiration or fermentation to generate ATP. While this process is less efficient than aerobic respiration, it allows cells to produce energy without oxygen.

a. Lactic Acid Fermentation

Location: Cytoplasm

Function: Converts pyruvate into lactic acid (lactate) while regenerating NAD⁺ from NADH.

Products: 2 ATP (per glucose molecule), 2 lactic acid molecules

Example: Occurs in muscle cells during intense exercise when oxygen supply is limited.

b. Alcohol Fermentation

Location: Cytoplasm

Function: Converts pyruvate into ethanol (alcohol) and CO_2, regenerating NAD^+ from NADH.

Products: 2 ATP (per glucose molecule), 2 ethanol molecules, 2 CO_2

Example: Used by yeast cells in the production of alcoholic beverages and bread.

6. Energy Yield and Efficiency

Aerobic Respiration: Produces approximately 36-38 ATP molecules per glucose molecule, making it highly efficient in energy production.

Anaerobic Respiration: Produces only 2 ATP molecules per glucose molecule, resulting in a lower yield of energy but allowing survival in anaerobic conditions.

7. Clinical and Practical Implications

Metabolic Disorders: Deficiencies in enzymes involved in cellular respiration can lead to metabolic disorders such as **Mitochondrial Diseases**, which affect energy production and can lead to muscle weakness, neurological problems, and other symptoms.

Cancer Metabolism: Cancer cells often exhibit altered metabolism, including increased glycolysis (Warburg effect) even in the presence of oxygen, which supports rapid cell proliferation.

Exercise and Energy: Understanding cellular respiration helps in optimizing athletic performance and recovery by knowing how different types of exercise affect aerobic and anaerobic energy production.

Cellular respiration is a vital process for producing ATP, the primary energy currency of the cell. It consists of glycolysis, the Krebs cycle, and oxidative phosphorylation, with aerobic respiration yielding the most ATP. In the absence of oxygen, cells rely on anaerobic respiration or fermentation. The efficiency of cellular respiration and its regulation are crucial for maintaining energy balance and supporting various physiological functions. Understanding these processes provides insights into energy metabolism, disease mechanisms, and practical applications in health and industry.

Photosynthesis

Photosynthesis is a fundamental biological process through which green plants, algae, and some bacteria convert light energy into chemical energy, storing it in the form of glucose. This process is crucial for the production of organic compounds and the release of oxygen into the atmosphere, making it vital for life on Earth.

1. Overview of Photosynthesis

Photosynthesis primarily occurs in the chloroplasts of plant cells and is composed of two main stages:

Light Reactions (Light-dependent Reactions)

Calvin Cycle (Light-independent Reactions or Dark Reactions)

a. General Equation of Photosynthesis

The overall chemical equation for photosynthesis is: $6 CO_2 + 6 H_2O + \text{light energy} \rightarrow C_6H_{12}O_6 + 6 O_2$

Reactants: Carbon dioxide (CO_2), water (H_2O), and light energy

Products: Glucose ($C_6H_{12}O_6$) and oxygen (O_2)

2. Light Reactions

Light Reactions occur in the thylakoid membranes of the chloroplasts and are responsible for converting light energy into chemical energy.

a. Photosynthetic Pigments

Chlorophyll: The primary pigment involved in photosynthesis, primarily chlorophyll a and b, absorbs light energy. Chlorophyll a is essential for the light reactions, while chlorophyll b helps in capturing light energy and passing it to chlorophyll a.

Accessory Pigments: Carotenoids (e.g., beta-carotene) and phycobilins capture additional light wavelengths and protect chlorophyll from damage.

b. Photochemical Phase

Photon Absorption: Chlorophyll absorbs photons, exciting electrons to a higher energy state.

Water Splitting (Photolysis): Water molecules are split into oxygen, protons, and electrons. The reaction is: $2H_2O \rightarrow 4H^+ + 4e^- + O_2$

Electron Transport Chain (ETC): Excited electrons are transferred through a series of proteins (photosystem II, plastoquinone, cytochrome b6f complex, plastocyanin, and photosystem I) in the thylakoid membrane. The energy from these electrons is used to pump protons into the thylakoid lumen, creating a proton gradient.

ATP and NADPH Formation: The proton gradient drives ATP synthesis via ATP synthase. Electrons reduce NADP$^+$ to NADPH. These products (ATP and NADPH) are used in the Calvin cycle.

c. Summary of Light Reactions

Products: Oxygen (O_2), ATP, NADPH

Location: Thylakoid membranes

3. Calvin Cycle

The **Calvin Cycle** (or Dark Reactions) occurs in the stroma of the chloroplasts and utilizes ATP and NADPH produced in the light reactions to convert carbon dioxide into glucose.

a. Phases of the Calvin Cycle

Carbon Fixation: Carbon dioxide is fixed into a 5-carbon sugar (ribulose bisphosphate, RuBP) by the enzyme RuBisCO, forming a 6-carbon intermediate that immediately splits into two 3-carbon molecules (3-phosphoglycerate, 3-PGA).

Reduction: ATP and NADPH are used to reduce 3-PGA into glyceraldehyde-3-phosphate (G3P), a 3-carbon sugar. This phase converts the fixed carbon into a more energy-rich form.

Regeneration: Some G3P molecules are used to regenerate RuBP, enabling the cycle to continue. This step requires additional ATP. The remaining G3P molecules are used to synthesize glucose and other carbohydrates.

b. Summary of the Calvin Cycle

Inputs: ATP, NADPH, CO_2

Outputs: Glucose ($C_6H_{12}O_6$), ADP, $NADP^+$

Location: Stroma of chloroplasts

4. Factors Affecting Photosynthesis

Several factors influence the rate of photosynthesis:

Light Intensity: As light intensity increases, the rate of photosynthesis also increases up to a certain point, beyond which it plateaus.

Carbon Dioxide Concentration: Higher CO_2 levels can enhance the rate of photosynthesis until other factors become limiting.

Temperature: Photosynthesis has an optimal temperature range. Extreme temperatures can denature enzymes or slow down reactions.

Water Availability: Water is a key reactant in photosynthesis; insufficient water can limit the process.

Chlorophyll Concentration: Adequate chlorophyll is necessary for capturing light energy.

5. Types of Photosynthesis

Different plants and organisms have adapted various photosynthetic pathways to optimize energy capture and use:

C3 Photosynthesis: The most common type, occurring in many plants. The Calvin cycle directly fixes CO_2 into a 3-carbon molecule.

C4 Photosynthesis: Found in plants like corn and sugarcane. This pathway involves an additional step where CO_2 is initially fixed into a 4-carbon compound (oxaloacetate) before entering the Calvin cycle, reducing photorespiration and increasing efficiency in hot, dry climates.

CAM Photosynthesis: Found in succulents and cacti. CAM plants open their stomata at night to fix CO_2 into malic acid, which is stored and then used during the day when the stomata are closed, minimizing water loss.

6. Practical and Ecological Significance

Oxygen Production: Photosynthesis is the primary source of atmospheric oxygen, essential for aerobic respiration in most living organisms.

Carbon Cycle: It plays a crucial role in the global carbon cycle by removing CO_2 from the atmosphere and converting it into organic matter.

Agriculture: Understanding photosynthesis is essential for improving crop yields and developing sustainable agricultural practices.

Climate Change: Plants' ability to sequester carbon helps mitigate climate change. Photosynthesis research informs strategies for enhancing carbon capture and reducing atmospheric CO_2 levels.

7. Clinical and Technological Applications

Artificial Photosynthesis: Research into artificial photosynthesis aims to replicate the natural process to produce clean energy and reduce CO_2 levels.

Bioengineering: Genetic modification of plants to enhance photosynthesis can lead to improved crop varieties and increased food production.

Environmental Monitoring: Photosynthesis data helps in assessing plant health and productivity, which can be used to monitor ecosystem changes and address environmental concerns.

Photosynthesis is a vital process that converts light energy into chemical energy, providing the foundation for life on Earth. It consists of light reactions, which generate ATP and NADPH, and the Calvin cycle, which synthesizes glucose from carbon dioxide. Various factors affect the efficiency of photosynthesis, and different types of photosynthetic pathways have evolved to optimize energy capture under different environmental conditions. Understanding photosynthesis is crucial for addressing global challenges related to energy, climate, and food security.

Cell Cycle and Mitosis

The **cell cycle** is a series of phases that a cell goes through as it grows and divides. It is essential for growth, development, and repair in multicellular organisms. **Mitosis** is a key part of the cell cycle where the cell's nucleus divides, resulting in two genetically identical daughter cells. This process ensures that each daughter cell receives an exact copy of the parent cell's DNA.

1. The Cell Cycle

The cell cycle is divided into two main phases: **interphase** and **mitotic phase** (M phase). Interphase is further divided into three stages: G1, S, and G2.

a. Interphase

Interphase is the phase where the cell grows, performs its functions, and prepares for division. It consists of three stages:

G1 Phase (Gap 1):

Activities: Cell growth, synthesis of proteins and organelles, and preparation for DNA replication.

Checkpoint: The cell checks for DNA damage, adequate size, and availability of nutrients. If conditions are favorable, the cell progresses to the S phase.

S Phase (Synthesis):

Activities: DNA replication occurs, where each chromosome is duplicated to form two sister chromatids. This ensures that each daughter cell will receive a complete set of chromosomes.

Checkpoint: The cell checks for errors in DNA replication and repairs them before proceeding to the next phase.

G2 Phase (Gap 2):

Activities: Further cell growth, synthesis of proteins necessary for mitosis, and preparation for cell division.

Checkpoint: The cell checks for DNA damage and ensures that all chromosomes have been replicated accurately. It also checks for adequate cell size and resources.

b. Mitotic Phase (M Phase)

The mitotic phase consists of mitosis (nuclear division) and cytokinesis (cytoplasmic division). It is the process by which a cell divides to form two genetically identical daughter cells.

Mitosis: Mitosis is divided into several stages:

Prophase:

Chromatin Condensation: Chromatin fibers condense into visible chromosomes, each consisting of two sister chromatids connected by a centromere.

Nuclear Envelope Breakdown: The nuclear envelope begins to disintegrate.

Mitotic Spindle Formation: The mitotic spindle, composed of microtubules, begins to form from centrosomes located at opposite poles of the cell.

Metaphase:

Chromosome Alignment: Chromosomes align along the metaphase plate (the cell's equatorial plane) with their centromeres attached to spindle fibers.

Anaphase:

Chromatid Separation: Sister chromatids are pulled apart toward opposite poles of the cell as spindle fibers shorten.

Cytokinesis Initiation: Cytokinesis, the division of the cytoplasm, begins during late anaphase.

Telophase:

Nuclear Envelope Reformation: The nuclear envelope reforms around each set of separated chromosomes.

Chromosome Decondensation: Chromosomes begin to uncoil back into chromatin.

Mitotic Spindle Disassembly: The mitotic spindle disassembles.

Cytokinesis:

Division of Cytoplasm: Cytokinesis completes the cell division process by dividing the cytoplasm and organelles into two daughter cells.

Animal Cells: Cytokinesis occurs through the formation of a cleavage furrow that pinches the cell membrane inwards.

Plant Cells: Cytokinesis occurs through the formation of a cell plate that develops into a new cell wall, separating the daughter cells.

2. Regulation of the Cell Cycle

The cell cycle is tightly regulated to ensure accurate cell division and prevent uncontrolled growth, which can lead to cancer.

a. Checkpoints

G1 Checkpoint: Ensures that the cell is ready to enter the S phase. Checks for DNA damage, cell size, and nutrient availability.

G2 Checkpoint: Ensures that DNA replication is complete and accurate before entering mitosis.

M Checkpoint (Spindle Assembly Checkpoint): Ensures that all chromosomes are properly attached to the spindle apparatus before proceeding to anaphase.

b. Regulatory Proteins

Cyclins: Proteins whose levels fluctuate throughout the cell cycle. They bind to cyclin-dependent kinases (CDKs) to activate them.

Cyclin-Dependent Kinases (CDKs): Enzymes that, when activated by cyclins, phosphorylate target proteins to drive the cell cycle forward.

Tumor Suppressors: Proteins such as p53 and Rb (retinoblastoma) that inhibit the cell cycle in response to DNA damage or other problems.

Oncogenes: Mutated genes that can promote uncontrolled cell division when activated.

3. Cell Cycle and Disease

Disruptions in cell cycle regulation can lead to diseases such as cancer:

Cancer: Caused by uncontrolled cell division due to mutations in genes regulating the cell cycle. Cancer cells often bypass checkpoints, leading to unregulated growth and the potential to form tumors.

Apoptosis: Programmed cell death that can be triggered by severe DNA damage or other cellular stress. Disruption of apoptosis can also contribute to cancer development.

4. Practical Applications

Understanding the cell cycle and mitosis has numerous practical implications:

Cancer Treatment: Targeting specific stages of the cell cycle or mitosis with drugs (e.g., taxanes, which inhibit spindle formation) can be effective in treating cancer.

Regenerative Medicine: Knowledge of cell division is crucial for developing stem cell therapies and tissue engineering.

Genetic Research: Studying cell division processes helps in understanding genetic disorders and developing gene therapies.

The cell cycle is a complex series of phases that ensure accurate cell division and replication. It consists of interphase (G1, S, G2) and the mitotic phase (mitosis and cytokinesis). Mitosis, the division of the nucleus, is divided into prophase, metaphase, anaphase, and telophase. Cytokinesis follows to divide the cytoplasm, resulting in two daughter cells. The cell cycle is regulated by checkpoints and regulatory proteins to maintain cellular integrity and prevent diseases such as cancer. Understanding these processes is essential for advancing medical treatments, particularly in cancer therapy and regenerative medicine.

Meiosis and Sexual Reproduction

Meiosis is a specialized type of cell division that reduces the chromosome number by half and results in the production of four non-identical gametes (sperm or eggs in animals, pollen or ovules in plants). This process is essential for sexual reproduction as it ensures genetic diversity and the maintenance of chromosome number across generations.

1. Overview of Meiosis

Meiosis occurs in two stages: **Meiosis I** and **Meiosis II**. Each stage consists of several phases similar to those in mitosis, but with key differences that contribute to genetic diversity.

a. Meiosis I

Meiosis I is known as the reduction division because it reduces the chromosome number by half.

Prophase I:

Chromosome Condensation: Chromosomes condense and become visible.

Synapsis: Homologous chromosomes pair up to form tetrads (a structure of four chromatids).

Crossing Over: Genetic material is exchanged between non-sister chromatids of homologous chromosomes at points called chiasmata. This process results in genetic recombination, increasing genetic diversity.

Nuclear Envelope Breakdown: The nuclear envelope disintegrates.

Spindle Formation: The mitotic spindle begins to form, and spindle fibers attach to the centromeres of the chromosomes.

Metaphase I:

Chromosome Alignment: Tetrads align along the metaphase plate with homologous chromosomes facing opposite poles of the cell.

Anaphase I:

Homologous Chromosome Separation: Homologous chromosomes are pulled to opposite poles of the cell. Unlike mitosis, the sister chromatids remain attached at this stage.

Telophase I:

Nuclear Envelope Formation: The nuclear envelope may reform around each set of chromosomes, and the chromosomes may de-condense.

Cytokinesis: The cytoplasm divides, resulting in two haploid cells, each with half the number of chromosomes.

b. Meiosis II

Meiosis II is similar to mitosis but occurs in haploid cells produced from Meiosis I.

Prophase II:

Chromosome Condensation: Chromosomes condense again.

Spindle Formation: The spindle apparatus reforms, and the nuclear envelope breaks down if it had reformed in telophase I.

Metaphase II:

Chromosome Alignment: Chromosomes align along the metaphase plate in each haploid cell. Sister chromatids are oriented toward opposite poles.

Anaphase II:

Sister Chromatid Separation: The sister chromatids are pulled apart to opposite poles of each cell.

Telophase II:

Nuclear Envelope Formation: The nuclear envelope reforms around each set of chromosomes.

Chromosome De-condensation: Chromosomes de-condense into chromatin.

Cytokinesis: The cytoplasm divides, resulting in four non-identical haploid cells, each with a single set of chromosomes.

2. Genetic Variation in Meiosis

Meiosis generates genetic diversity through several mechanisms:

Crossing Over:

Occurs in Prophase I: Homologous chromosomes exchange segments, resulting in recombinant chromosomes with new combinations of alleles.

Independent Assortment:

Occurs in Metaphase I: The orientation of homologous chromosome pairs at the metaphase plate is random, leading to various possible combinations of chromosomes in gametes.

Random Fertilization:

Post-meiosis: The combination of gametes during fertilization is random, contributing further to genetic diversity in the offspring.

3. Sexual Reproduction

Sexual reproduction involves the fusion of two gametes (sperm and egg in animals, pollen and ovule in plants) to form a zygote. This process combines genetic material from two parents, resulting in offspring with unique genetic combinations.

a. Gametogenesis

Spermatogenesis: The process by which male gametes (sperm) are produced. It occurs in the testes and results in four functional sperm cells from one precursor cell.

Oogenesis: The process by which female gametes (eggs or ova) are produced. It occurs in the ovaries and results in one functional egg cell and three polar bodies (which are typically not functional).

b. Fertilization

Fusion of Gametes: During fertilization, a sperm cell fuses with an egg cell to form a zygote, which then undergoes mitotic divisions to develop into a new organism.

Genetic Variation: Fertilization combines the genetic material from two parents, resulting in genetic variation among offspring.

4. Differences Between Mitosis and Meiosis

Purpose:

Mitosis: Produces two genetically identical diploid cells for growth, repair, and asexual reproduction.

Meiosis: Produces four genetically diverse haploid gametes for sexual reproduction.

Chromosome Number:

Mitosis: Maintains the chromosome number (diploid to diploid).

Meiosis: Reduces the chromosome number by half (diploid to haploid).

Genetic Variation:

Mitosis: Results in genetically identical daughter cells.

Meiosis: Introduces genetic variation through crossing over and independent assortment.

5. Clinical and Practical Implications

Genetic Disorders: Errors in meiosis, such as nondisjunction (failure of chromosomes to separate properly), can lead to genetic disorders like Down syndrome (trisomy 21) and Turner syndrome.

Assisted Reproductive Technologies (ART): Techniques like in vitro fertilization (IVF) rely on understanding meiosis and gametogenesis to help individuals conceive.

Evolutionary Biology: The mechanisms of genetic variation introduced by meiosis contribute to evolutionary processes and the adaptability of species.

Meiosis is a crucial process in sexual reproduction that reduces the chromosome number by half and introduces genetic diversity through crossing over and independent assortment. It consists of two stages: Meiosis I (reduction division) and Meiosis II (equational division). Understanding meiosis is essential for comprehending genetic variation, reproduction, and the basis of many genetic disorders. Its mechanisms have broad implications in medicine, evolutionary biology, and reproductive technologies.

Part III: Genetics

Chapter 7: DNA and RNA

Structure of DNA and RNA

The molecules **DNA** (Deoxyribonucleic Acid) and **RNA** (Ribonucleic Acid) are essential for the storage, transmission, and expression of genetic information in all living organisms. Their structures enable them to perform their functions efficiently, with **DNA** being the hereditary material and **RNA** playing various roles in protein synthesis and gene regulation.

1. Structure of DNA

DNA is a double-stranded molecule composed of nucleotides, which are the building blocks of both DNA and RNA. A nucleotide consists of three components:

A nitrogenous base: Adenine (A), Thymine (T), Cytosine (C), or Guanine (G)

A pentose sugar: In DNA, the sugar is deoxyribose.

A phosphate group: Forms the backbone of the DNA molecule.

a. Double-Helical Structure

The **double-helix model** of DNA was first proposed by James Watson and Francis Crick in 1953, with the assistance of Rosalind Franklin's X-ray diffraction data. This model describes the structural features of DNA:

Two Strands: DNA consists of two long strands of nucleotides running in opposite directions (antiparallel), forming a ladder-like structure twisted into a helix.

Antiparallel Orientation: One strand runs in the 5' to 3' direction, while the other runs in the 3' to 5' direction, which is crucial for DNA replication and repair.

Sugar-Phosphate Backbone: The sugar (deoxyribose) and phosphate groups of each nucleotide form the backbone of each strand, connected by **phosphodiester bonds**. These bonds link the 5' carbon of one sugar to the 3' carbon of the next sugar.

Base Pairing: The nitrogenous bases project inward from the backbone and pair with complementary bases from the opposite strand, forming the "rungs" of the ladder.

Complementary Base Pairing: Adenine (A) pairs with Thymine (T) through **two hydrogen bonds**, and Cytosine (C) pairs with Guanine (G) through **three hydrogen bonds**. This pairing is known as **Chargaff's Rule**.

Helical Shape: The structure twists into a right-handed double helix, with approximately 10 base pairs per turn of the helix. This shape helps compact the DNA molecule to fit into the cell's nucleus and provides stability.

b. Major and Minor Grooves

In the double helix, the twisting of the strands creates **major** and **minor grooves**. These grooves are important because they allow proteins, such as transcription factors, to interact with specific sequences of DNA, facilitating gene regulation, replication, and repair.

c. Supercoiling and Chromatin

Supercoiling: In cells, DNA is further compacted by supercoiling, which occurs when the DNA helix is overwound or underwound. This compact structure is essential for fitting large amounts of DNA into the cell nucleus.

Chromatin: In eukaryotes, DNA is packaged into chromatin, where it is wrapped around histone proteins, forming nucleosomes. This organization allows the DNA to be more accessible for processes like transcription and replication.

2. Structure of RNA

RNA is a single-stranded nucleic acid, but it shares some similarities with DNA. It also consists of nucleotides, but with some key differences.

a. RNA Nucleotide Components

Nitrogenous Bases: RNA contains four bases, but instead of Thymine (T), RNA uses **Uracil (U)**. The other three bases are Adenine (A), Cytosine (C), and Guanine (G).

Base Pairing: In RNA, Adenine pairs with Uracil (A-U), and Cytosine pairs with Guanine (C-G) when RNA forms double-stranded regions.

Pentose Sugar: RNA contains **ribose** instead of deoxyribose. The presence of an additional hydroxyl (-OH) group on the 2' carbon of the sugar makes RNA more chemically reactive and less stable than DNA.

Phosphate Group: Like DNA, RNA has a phosphate backbone, which forms **phosphodiester bonds** between nucleotides.

b. Single-Stranded Structure

Unlike DNA, RNA is typically single-stranded, though it can fold into complex secondary and tertiary structures by forming intramolecular base pairs. These structures are important for RNA's diverse functions in the cell.

c. Types of RNA

There are several types of RNA, each with unique functions in the cell:

Messenger RNA (mRNA): Carries genetic information from DNA to the ribosome, where proteins are synthesized.

Transfer RNA (tRNA): Brings the appropriate amino acids to the ribosome during protein synthesis, matching them with the codons on the mRNA.

Ribosomal RNA (rRNA): Forms the core of the ribosome's structure and catalyzes protein synthesis.

Other Functional RNAs:

MicroRNA (miRNA) and **small interfering RNA (siRNA)**: Involved in gene regulation by silencing specific mRNA molecules.

Small nuclear RNA (snRNA): Plays a role in RNA splicing, a process where introns are removed from pre-mRNA.

3. Key Differences Between DNA and RNA

Feature	DNA	RNA
Strands	Double-stranded	Single-stranded
Sugar	Deoxyribose	Ribose
Bases	Adenine, Thymine, Cytosine, Guanine	Adenine, Uracil, Cytosine, Guanine
Stability	More stable (due to double-strand and fewer reactive -OH groups)	Less stable (due to single strand and reactive ribose)
Function	Long-term storage of genetic information	Various roles in protein synthesis and gene regulation

4. Functions of DNA and RNA

a. Functions of DNA

Genetic Information Storage: DNA carries the genetic blueprint of an organism and is passed from generation to generation. It determines the traits and characteristics of an organism by encoding proteins and regulatory elements.

Replication: DNA can replicate itself before cell division, ensuring that each daughter cell receives an exact copy of the genetic material.

Transcription: DNA serves as a template for the synthesis of RNA through the process of transcription, which is the first step in gene expression.

b. Functions of RNA

Protein Synthesis: RNA is central to the process of translating the genetic code in DNA into proteins. mRNA carries the code, tRNA brings amino acids, and rRNA helps catalyze the formation of peptide bonds in the ribosome.

Gene Regulation: RNA molecules such as miRNA and siRNA regulate gene expression by degrading mRNA or inhibiting its translation.

Catalysis: Certain RNA molecules, known as ribozymes, have catalytic properties and can participate in biochemical reactions, such as RNA splicing.

5. DNA and RNA in Genetic Processes

Replication: DNA replication is the process of copying the DNA before cell division. It ensures that each daughter cell inherits a full set of chromosomes. DNA polymerase is the key enzyme involved.

Transcription: Transcription is the process by which RNA is synthesized from a DNA template. RNA polymerase binds to the DNA and creates a complementary RNA strand.

Translation: Translation is the process where the sequence of mRNA is used to assemble amino acids into proteins. This occurs in the ribosome, with the help of tRNA and rRNA.

DNA and RNA are nucleic acids essential to life, each with distinct structures and roles. DNA is a double-stranded helix responsible for long-term genetic storage and transmission. RNA, a single-stranded molecule, plays versatile roles in protein synthesis and gene regulation. DNA's stable structure allows it to store genetic information, while RNA's structure enables its function in gene expression and regulation. Understanding the structures of DNA and RNA is crucial to comprehending how genetic information is passed, expressed, and regulated in living organisms.

DNA Replication

DNA replication is the biological process by which a cell duplicates its DNA, producing two identical copies from a single original DNA molecule. This process is crucial for cell division, ensuring that each daughter cell inherits an exact copy of the parent cell's genetic material. DNA replication is a highly regulated and precise mechanism that involves multiple enzymes and steps.

1. Overview of DNA Replication

DNA replication follows a **semi-conservative** model, which was confirmed by the Meselson-Stahl experiment in 1958. In this model, each new DNA molecule consists of one **original (parental) strand** and one **newly synthesized strand**. This process occurs in the **S phase** of the cell cycle, prior to mitosis or meiosis, and involves the following key features:

Semi-conservative: Each daughter DNA molecule retains one original parental strand and incorporates one new strand.

Bidirectional: Replication begins at specific locations (origins of replication) and proceeds in two directions along the DNA.

Semi-discontinuous: One strand (the leading strand) is synthesized continuously, while the other (the lagging strand) is synthesized in short fragments.

2. Key Enzymes and Proteins Involved in DNA Replication

Several enzymes and proteins work together to ensure accurate and efficient DNA replication:

Helicase: Unwinds the double-stranded DNA at the replication fork, creating two single strands that serve as templates for replication.

Single-Strand Binding Proteins (SSBs): Bind to the separated strands to prevent them from re-annealing and protect the single strands from degradation.

Topoisomerase: Relieves the torsional strain generated by the unwinding of the DNA helix by creating temporary breaks in the DNA strands and then rejoining them.

Primase: Synthesizes a short RNA primer that provides a starting point for DNA synthesis. DNA polymerases can only add nucleotides to an existing strand, so the primer is essential.

DNA Polymerase: The main enzyme responsible for adding nucleotides to the growing DNA chain, using the parental strand as a template. Different types of DNA polymerase exist in prokaryotes and eukaryotes.

Prokaryotes: DNA polymerase III is the primary enzyme for DNA synthesis.

Eukaryotes: DNA polymerases α, δ, and ε play major roles in replication.

DNA Ligase: Seals the gaps between Okazaki fragments (short DNA sequences synthesized on the lagging strand), creating a continuous DNA strand.

Sliding Clamp (PCNA in eukaryotes): Ensures that DNA polymerase stays attached to the DNA template, improving processivity (the number of nucleotides added before the polymerase dissociates).

3. Steps of DNA Replication

a. Initiation

DNA replication begins at specific sites called **origins of replication**. These are particular sequences in the DNA where replication is initiated. In **prokaryotes** (which have circular DNA), there is typically a single origin, while **eukaryotes** (with linear chromosomes) have multiple origins of replication.

Origin Recognition: In eukaryotes, the **origin recognition complex (ORC)** binds to the origin of replication, marking it as the starting point for replication. In prokaryotes, proteins like **DnaA** perform a similar function.

Unwinding the DNA: **Helicase** is recruited to the origin of replication and begins unwinding the DNA helix, creating two single strands. **Single-strand binding proteins (SSBs)** bind to the exposed single strands to stabilize them and prevent re-annealing.

Formation of the Replication Fork: As helicase unwinds the DNA, two replication forks form, where replication machinery can begin synthesizing new strands. The area where the DNA is actively being unwound and replicated is known as the **replication bubble**.

b. Elongation

The elongation phase is where the actual synthesis of new DNA strands occurs. This phase differs slightly for the **leading** and **lagging** strands due to the antiparallel nature of DNA.

Leading Strand Synthesis:

DNA synthesis proceeds **continuously** in the 5' to 3' direction (toward the replication fork).

Primase synthesizes a short RNA primer, providing a 3' hydroxyl (-OH) group for DNA polymerase to begin adding nucleotides.

DNA polymerase III (in prokaryotes) or **DNA polymerase δ** (in eukaryotes) extends the primer by adding nucleotides complementary to the template strand, moving continuously toward the replication fork.

Lagging Strand Synthesis:

DNA synthesis proceeds **discontinuously** in the opposite direction (away from the replication fork).

Since DNA polymerase can only synthesize in the 5' to 3' direction, **primase** must lay down multiple RNA primers along the lagging strand.

DNA polymerase III (in prokaryotes) or **DNA polymerase δ/α** (in eukaryotes) synthesizes short segments of DNA called **Okazaki fragments**, which are extended from the RNA primers.

As the replication fork moves, additional primers are laid down, and more Okazaki fragments are synthesized.

c. Primer Removal and Ligation

Once the DNA strands are synthesized, the RNA primers must be removed and replaced with DNA, and the Okazaki fragments must be joined together to form a continuous strand.

Primer Removal:

In prokaryotes, **DNA polymerase I** removes the RNA primers and replaces them with DNA.

In eukaryotes, **RNase H** degrades the RNA primers, and DNA polymerase δ fills in the gaps with DNA.

Ligation:

After the RNA primers are replaced with DNA, there are still gaps between the Okazaki fragments.

DNA ligase seals these gaps by creating a phosphodiester bond between the 3' hydroxyl end of one fragment and the 5' phosphate end of the next fragment, ensuring a continuous DNA strand.

d. Termination

Replication ends when the replication forks meet, and the entire DNA molecule has been copied. In **prokaryotes**, this occurs when the replication forks meet at a specific termination site. In **eukaryotes**, termination is more complex due to the linear nature of chromosomes.

Prokaryotic Termination: Termination occurs when two replication forks meet at a region known as the ter site. Special termination proteins, such as Tus, help stop the movement of the replication machinery.

Eukaryotic Termination: Eukaryotic chromosomes have **telomeres**, repetitive sequences at their ends that prevent the loss of important genetic information during replication. The enzyme **telomerase** adds these repetitive sequences to the ends of chromosomes to protect them.

4. Accuracy of DNA Replication: Proofreading and Repair

DNA replication is highly accurate due to the proofreading ability of DNA polymerases and the presence of various repair mechanisms.

Proofreading by DNA Polymerase:

DNA polymerase has a **3' to 5' exonuclease activity**, which allows it to remove incorrectly paired nucleotides immediately after they are added. This reduces the error rate during DNA replication.

Mismatch Repair: After replication, any errors that escape proofreading can be corrected by the mismatch repair system, which recognizes and removes mismatched bases and replaces them with the correct ones.

5. Differences Between Prokaryotic and Eukaryotic DNA Replication

Feature	Prokaryotic DNA Replication	Eukaryotic DNA Replication
Origins of Replication	Single origin of replication	Multiple origins of replication
Replication Speed	Faster due to smaller genome size	Slower due to larger genome size
Primase Activity	Single primase for both leading and lagging strand synthesis	Complex primase (part of DNA polymerase α)
Polymerase Types	DNA polymerase III (leading and lagging strand synthesis)	DNA polymerases α, δ, and ε involved in replication
Termination	Circular chromosomes allow termination at specific sites	Linear chromosomes require telomerase to maintain ends

6. Telomeres and Telomerase in Eukaryotic DNA Replication

Telomeres are repetitive DNA sequences at the ends of linear chromosomes that protect the chromosome from deterioration or fusion with neighboring chromosomes. With each round of DNA replication, the ends of linear chromosomes shorten slightly because DNA polymerase cannot fully replicate the extreme 3' end of the lagging strand.

To prevent the loss of essential genetic material, the enzyme **telomerase** adds extra nucleotide sequences to the ends of telomeres. Telomerase is especially active in germ cells, stem cells, and cancer cells, which require continuous division. In most somatic cells, telomerase activity is low, leading to gradual telomere shortening over time, which is associated with aging and cellular senescence.

DNA replication is a highly coordinated and regulated process essential for cell division. It ensures the faithful duplication of genetic material, allowing cells to pass on genetic information to their offspring. The semi-conservative nature of replication, the bidirectional replication fork movement, and the involvement of numerous enzymes ensure that the process is both accurate and efficient.

Transcription and Translation

Transcription and translation are two essential processes in the flow of genetic information from DNA to functional proteins, known as the **central dogma of molecular biology**. **Transcription** is the process by which RNA is synthesized from a DNA template, while **translation** is the process by which the RNA message is decoded to synthesize proteins.

1. Overview of Transcription

Transcription is the process in which a specific segment of DNA is copied into **messenger RNA (mRNA)** by the enzyme RNA polymerase. This is the first step of gene expression, as the information in DNA is transcribed into a form that can be used to produce proteins.

a. The Steps of Transcription

Transcription occurs in three stages: **initiation**, **elongation**, and **termination**.

1. Initiation

Transcription begins at specific DNA sequences called **promoters**. A promoter is a region of DNA that signals the RNA polymerase where to begin transcription.

In **prokaryotes**, the promoter usually contains a **TATA box** and other conserved sequences that help RNA polymerase recognize the starting point.

In **eukaryotes**, transcription factors are required to help RNA polymerase bind to the promoter.

Once bound to the promoter, RNA polymerase separates the DNA strands, creating a **transcription bubble**.

2. Elongation

During elongation, RNA polymerase reads the DNA template strand in the **3' to 5' direction** and synthesizes RNA in the **5' to 3' direction**.

RNA polymerase adds complementary **ribonucleotides** (A, U, G, C) to the growing RNA chain. Unlike DNA, RNA contains **uracil (U)** instead of thymine (T).

As RNA polymerase moves along the DNA, the DNA strands re-form into their double helix, and the RNA strand elongates.

3. Termination

When RNA polymerase reaches a specific sequence called a **terminator**, the enzyme stops transcribing.

In **prokaryotes**, termination is usually signaled by a specific sequence, such as a **hairpin loop** in the RNA that causes the polymerase to disassociate.

In **eukaryotes**, termination is more complex and involves the addition of a **polyadenylation signal** (AAUAAA), leading to the release of the newly synthesized RNA strand.

b. Post-Transcriptional Modifications (Eukaryotes)

In **eukaryotes**, the primary RNA transcript, called **pre-mRNA**, undergoes several modifications before it becomes mature **mRNA** capable of being translated into a protein.

5' Cap Addition: A modified guanine nucleotide (7-methylguanosine) is added to the 5' end of the mRNA. This **5' cap** protects the RNA from degradation and is important for ribosome binding during translation.

Polyadenylation: A tail of 100-200 adenine nucleotides (**poly-A tail**) is added to the 3' end of the mRNA. This tail enhances the stability of the mRNA and facilitates its export from the nucleus to the cytoplasm.

RNA Splicing: Eukaryotic genes contain both coding regions (**exons**) and non-coding regions (**introns**). Before the mRNA can be translated, the introns must be removed in a process called **splicing**. The **spliceosome**, a complex of proteins and small nuclear RNAs (snRNAs), excises introns and joins exons together to form a continuous coding sequence.

After these modifications, the mature mRNA is exported from the nucleus into the cytoplasm, where it can be translated into a protein.

c. Differences in Transcription Between Prokaryotes and Eukaryotes

Feature	Prokaryotic Transcription	Eukaryotic Transcription
Location	Cytoplasm	Nucleus
RNA Polymerase Types	One type of RNA polymerase	Three types: RNA polymerase I, II, III
Post-Transcriptional Modifications	None (no capping, poly-A tail, or splicing)	5' capping, poly-A tail addition, splicing
Coupling with Translation	Transcription and translation occur simultaneously	Transcription and translation are separated spatially
Termination	Simpler mechanisms (e.g., rho-dependent and rho-independent)	Complex termination involving polyadenylation signals

2. Overview of Translation

Translation is the process by which the sequence of nucleotides in the mRNA is decoded to assemble a chain of amino acids, forming a protein. This process occurs in the ribosome, a molecular machine found in the cytoplasm.

a. The Genetic Code

The mRNA sequence is read in groups of three nucleotides called codons. Each codon specifies a particular amino acid, according to the genetic code. The genetic code is:

Universal: It is shared by almost all organisms.

Redundant: Several codons can code for the same amino acid.

Unambiguous: Each codon specifies only one amino acid.

There are 64 codons in total, which include:

61 codons that specify amino acids.

3 stop codons (UAA, UAG, UGA) that signal the end of translation.

1 start codon (AUG), which codes for the amino acid methionine and initiates translation.

b. Transfer RNA (tRNA)

tRNA molecules are essential for translation as they carry amino acids to the ribosome. Each tRNA has:

An anticodon: A set of three nucleotides complementary to the mRNA codon.

An attached amino acid: The specific amino acid that the tRNA delivers to the growing polypeptide chain.

tRNA molecules match their anticodons with the appropriate codons on the mRNA to ensure the correct amino acids are incorporated into the protein.

c. The Ribosome

The ribosome is the cellular structure that facilitates translation. It consists of two subunits:

Small subunit: Binds to the mRNA and ensures proper alignment of the tRNA with the codon.

Large subunit: Contains the peptidyl transferase activity that forms peptide bonds between amino acids.

The ribosome has three functional sites:

A (aminoacyl) site: Holds the tRNA carrying the next amino acid to be added to the chain.

P (peptidyl) site: Holds the tRNA carrying the growing polypeptide chain.

E (exit) site: The site where the empty tRNA leaves the ribosome after its amino acid has been added.

d. The Stages of Translation

Translation occurs in three stages: initiation, elongation, and termination.

1. Initiation

The small ribosomal subunit binds to the 5' end of the mRNA.

The initiator tRNA, carrying methionine and containing the anticodon for the start codon (AUG), binds to the start codon on the mRNA.

The large ribosomal subunit then joins the complex, forming a complete ribosome ready for translation.

2. Elongation

During elongation, amino acids are added one by one to the growing polypeptide chain.

A charged tRNA (tRNA with an amino acid attached) enters the A site of the ribosome, matching its anticodon with the mRNA codon.

The ribosome catalyzes the formation of a peptide bond between the amino acid on the tRNA in the P site and the new amino acid in the A site.

The ribosome then moves down the mRNA by one codon (translocation), shifting the tRNA in the A site to the P site and the tRNA in the P site to the E site, where it is released.

This process repeats, with new tRNAs bringing amino acids to the A site, and the polypeptide chain growing in length.

3. Termination

Translation ends when the ribosome encounters a stop codon (UAA, UAG, UGA) on the mRNA.

Instead of a tRNA, a release factor binds to the stop codon in the A site.

The release factor triggers the ribosome to release the completed polypeptide chain.

The ribosome subunits then dissociate, and translation is complete.

e. Post-Translational Modifications

After translation, proteins often undergo further modifications to become fully functional. These include:

Folding: The protein folds into its specific three-dimensional shape, often with the help of molecular chaperones.

Cleavage: Some proteins require cleavage of certain amino acid sequences (e.g., removal of the initiator methionine).

Chemical Modifications: Proteins may undergo phosphorylation, glycosylation, methylation, or acetylation to regulate their activity or stability.

3. Differences Between Transcription and Translation in Prokaryotes and Eukaryotes

Aspect	Prokaryotes	Eukaryotes
Location of Transcription	Cytoplasm	Nucleus
Location of Translation	Cytoplasm (ribosomes)	Cytoplasm (ribosomes in rough ER or free in cytoplasm)
mRNA Processing	No mRNA processing (no splicing, capping, or polyadenylation)	mRNA undergoes splicing, capping, and poly-A tail addition
Timing	Transcription and translation can occur simultaneously	Transcription and translation are separate processes

4. Summary

Transcription is the process where DNA is transcribed into mRNA, which serves as a template for protein synthesis.

Translation is the process where the mRNA is decoded in the ribosome to form a protein, with the help of tRNA and ribosomal subunits.

Both processes are critical for gene expression and are regulated differently in prokaryotes and eukaryotes.

Chapter 8: Mendelian Genetics

Gregor Mendel's Experiments

Gregor Mendel, often referred to as the "Father of Genetics," laid the foundation of modern genetics through his pioneering work on **inheritance**. His experiments with **pea plants** (Pisum sativum) between 1856 and 1863 provided the first significant insight into how traits are passed from one generation to the next. Although Mendel's work went unnoticed during his lifetime, it became the cornerstone of modern genetics after its rediscovery in the early 20th century.

1. Background: Why Pea Plants?

Mendel selected pea plants for his experiments for several reasons:

Clear, distinct traits: Pea plants exhibited obvious, contrasting traits (e.g., tall vs. short, yellow seeds vs. green seeds).

Controlled pollination: Pea plants could be easily cross-pollinated or self-pollinated, allowing Mendel to control the breeding process.

Rapid generation time: Pea plants grew quickly, allowing Mendel to observe results over multiple generations in a short period.

Large number of offspring: Pea plants produced a large number of seeds, providing Mendel with statistically meaningful data.

Mendel focused on **seven traits** in pea plants, each of which had two distinct forms. These traits were:

Seed shape: Round or wrinkled

Seed color: Yellow or green

Flower color: Purple or white

Pod shape: Inflated or constricted

Pod color: Green or yellow

Flower position: Axial (along the stem) or terminal (at the tip)

Plant height: Tall or short

2. Mendel's Experimental Design

Mendel conducted his experiments in a systematic and organized way, beginning with **pure-breeding** (or true-breeding) plants. Pure-breeding plants consistently produced offspring with the same traits when self-pollinated, suggesting they carried two identical versions (alleles) of the trait.

a. Monohybrid Cross: One Trait at a Time

Mendel's first series of experiments involved **monohybrid crosses**, where he focused on the inheritance of a single trait. For example, he crossed plants that produced **yellow seeds** with those that produced **green seeds**.

Step 1: Parental Generation (P)

Mendel began by cross-pollinating pure-breeding plants with contrasting traits. For example, he crossed pure-breeding yellow-seeded plants (YY) with pure-breeding green-seeded plants (yy).

Step 2: First Filial Generation (F1)

The offspring of this cross (F1 generation) all displayed the **dominant trait**—in this case, yellow seeds. None of the F1 plants had green seeds, indicating that the yellow trait dominated over the green trait.

Step 3: Second Filial Generation (F2)

Mendel then allowed the F1 plants to **self-pollinate**. The resulting F2 generation exhibited a **3:1 ratio** of yellow seeds to green seeds. This suggested that the green trait had not disappeared but was simply masked in the F1 generation.

b. Key Observations from Monohybrid Crosses

Dominance: Some traits (such as yellow seed color) dominate over others (green seed color). The dominant trait is expressed in the F1 generation, while the recessive trait is masked but can reappear in the F2 generation.

Segregation: Mendel proposed that each organism carries **two alleles** (one from each parent) for a given trait. These alleles segregate, or separate, during **gamete formation** (meiosis), and each gamete carries only one allele. This is known as Mendel's **Law of Segregation**.

In the case of seed color:

The **yellow allele (Y)** is dominant.

The **green allele (y)** is recessive.

F1 generation (Yy) plants express the dominant yellow color, but in the F2 generation, the recessive green color reappears in a **3:1 ratio** (3 yellow-seeded plants for every 1 green-seeded plant).

c. Dihybrid Cross: Two Traits at a Time

Mendel also performed **dihybrid crosses**, where he examined the inheritance of two traits simultaneously. For example, he crossed plants with **yellow, round seeds** (YYRR) with plants that had **green, wrinkled seeds** (yyrr).

Step 1: Parental Generation (P)

Mendel crossed pure-breeding plants for two traits: yellow, round seeds (YYRR) with green, wrinkled seeds (yyrr).

Step 2: First Filial Generation (F1)

All F1 offspring had yellow, round seeds (YyRr), showing that yellow and round were dominant traits.

Step 3: Second Filial Generation (F2)

When the F1 plants self-pollinated, Mendel observed four phenotypes in the F2 generation: yellow/round, yellow/wrinkled, green/round, and green/wrinkled.

These phenotypes occurred in a **9:3:3:1 ratio**, which suggested that the traits for seed color and seed shape were inherited independently of each other.

d. Key Observations from Dihybrid Crosses

Independent Assortment: The inheritance of one trait does not affect the inheritance of another. This is Mendel's **Law of Independent Assortment**, which states that alleles for different traits assort independently during gamete formation. This law applies only to genes located on different chromosomes or far apart on the same chromosome.

3. Mendel's Laws of Inheritance

Mendel's experiments led to the formulation of two fundamental laws of genetics:

a. Law of Segregation

Each organism carries two alleles for each trait, one inherited from each parent. During the formation of gametes (sperm or eggs), the alleles separate, so that each gamete carries only one allele for each trait. This explains why recessive traits can reappear in subsequent generations after being masked in the first.

b. Law of Independent Assortment

Alleles for different traits are inherited independently of each other. This means that the inheritance of one trait (e.g., seed color) does not influence the inheritance of another trait (e.g., seed shape). This principle is true for genes that are on different chromosomes or far apart on the same chromosome.

4. Significance of Mendel's Work

Mendel's experiments were groundbreaking for several reasons:

Systematic Approach: Mendel's careful observation, use of mathematical ratios, and methodical approach distinguished his work from previous studies of inheritance, which were more anecdotal and less precise.

Predictability: Mendel showed that inheritance follows specific rules and that it is possible to predict the outcome of genetic crosses.

Foundation of Genetics: His discovery of dominant and recessive traits, segregation, and independent assortment laid the groundwork for modern genetics, leading to the discovery of chromosomes, DNA, and the mechanisms of heredity.

5. Rediscovery and Impact

Mendel's work remained largely unknown until 1900, when three botanists—**Hugo de Vries**, **Carl Correns**, and **Erich von Tschermak**—independently rediscovered his findings. Mendel's principles became the foundation for the rapidly growing field of genetics, influencing studies on chromosomes, gene mapping, and molecular biology.

6. Exceptions and Extensions of Mendelian Genetics

While Mendel's laws apply to many traits, modern genetics has shown that some traits do not follow simple Mendelian inheritance patterns. Examples include:

Incomplete Dominance: In some cases, the heterozygous phenotype is intermediate between the two homozygous phenotypes (e.g., red and white flowers producing pink offspring).

Codominance: Both alleles are fully expressed in the heterozygous condition (e.g., AB blood type, where both A and B alleles are expressed).

Polygenic Inheritance: Traits like height and skin color are controlled by multiple genes, leading to a wide range of phenotypes.

Linked Genes: Genes that are located close together on the same chromosome tend to be inherited together, violating the principle of independent assortment.

Gregor Mendel's meticulous experiments on pea plants provided the first insight into the rules governing inheritance, forming the basis of classical genetics. His discovery of dominant and recessive alleles, and the formulation of the laws of segregation and independent assortment, were monumental in understanding how traits are passed from one generation to the next. Although Mendel worked in the 19th century, his contributions continue to influence genetic research and our understanding of biology today.

Laws of Inheritance

Gregor Mendel's work on pea plants in the mid-1800s led to the formulation of the **laws of inheritance**, which describe the fundamental principles by which traits are passed from parents to offspring. These laws, based on Mendel's observations of genetic crosses, form the foundation of classical genetics. There are two primary laws Mendel identified: the **Law of Segregation** and the **Law of Independent Assortment**. Later research has shown that while these laws apply to many traits, they are not universally applicable to all inheritance patterns.

1. Law of Segregation

The **Law of Segregation** is Mendel's first law, and it describes how alleles for a single trait separate during the formation of gametes (sperm and egg cells) so that each gamete receives only one allele for each trait. This law can be summarized as follows:

a. Definition

Each individual has two alleles for each gene (one inherited from each parent), but these alleles segregate (separate) during gamete formation, resulting in gametes that carry only one allele for each gene. When gametes fuse during fertilization, the offspring inherit one allele from each parent, thus restoring the pair of alleles.

b. Experiment Basis

This law was derived from Mendel's **monohybrid crosses**—experiments in which he tracked the inheritance of a single trait, such as seed color. For example:

P Generation: Mendel crossed pure-breeding pea plants with yellow seeds (YY) with those that had green seeds (yy). The plants in the **F1 generation** were all yellow-seeded (Yy), demonstrating that yellow was dominant over green.

F2 Generation: When the F1 plants self-pollinated, the F2 generation showed a **3:1 ratio** of yellow to green seeds. This revealed that the green trait had not disappeared but had been masked in the F1 generation. It reappeared because the Y and y alleles segregated during gamete formation in the F1 plants, and some offspring inherited two copies of the recessive y allele.

c. Genetic Explanation

Each gene exists in different forms, known as **alleles**. For example, in pea plants, the gene for seed color has two alleles: one for yellow seeds (Y) and one for green seeds (y). Each parent has two alleles for each gene, and these alleles segregate during the formation of gametes so that each gamete carries only one allele.

During **meiosis**, the process of cell division that produces gametes, the homologous chromosomes (which carry alleles for each gene) are separated into different gametes.

In the case of a heterozygous plant (Yy), half the gametes will carry the Y allele, and half will carry the y allele. These gametes combine randomly with those of another plant during fertilization, restoring the pair of alleles in the offspring.

d. Summary of the Law of Segregation

Each individual has two alleles for a particular gene (one from each parent).

During gamete formation, these alleles separate, so each gamete carries only one allele for each gene.

Offspring inherit one allele from each parent, thus restoring the pair of alleles.

2. Law of Independent Assortment

The **Law of Independent Assortment** is Mendel's second law, and it explains how alleles of different genes are distributed independently of one another during the formation of gametes. This law applies when genes are located on different chromosomes or far apart on the same chromosome, allowing them to assort independently.

a. Definition

Alleles for different genes assort independently of each other during gamete formation. This means that the inheritance of one trait (e.g., seed color) does not affect the inheritance of another trait (e.g., seed shape), as long as the genes controlling these traits are located on different chromosomes.

b. Experiment Basis

Mendel derived this law from his **dihybrid crosses**, where he examined the inheritance of two traits at the same time, such as seed color (yellow or green) and seed shape (round or wrinkled). For example:

P Generation: Mendel crossed plants that were pure-breeding for yellow, round seeds (YYRR) with plants that had green, wrinkled seeds (yyrr).

F1 Generation: All F1 offspring were heterozygous (YyRr) and displayed the dominant traits—yellow seeds and round seeds.

F2 Generation: When the F1 plants self-pollinated, Mendel observed four different combinations of traits in the F2 generation: yellow/round, yellow/wrinkled, green/round, and green/wrinkled. These traits appeared in a **9:3:3:1 ratio**, which indicated that the alleles for seed color and seed shape had assorted independently.

c. Genetic Explanation

The Law of Independent Assortment can be explained by the behavior of chromosomes during meiosis. Each pair of homologous chromosomes (and the alleles they carry) lines up randomly along the metaphase plate during **meiosis I**. As a result, the alleles for one gene are inherited independently of the alleles for other genes.

For example, in a plant that is heterozygous for both seed color (Yy) and seed shape (Rr), the Y and y alleles assort independently of the R and r alleles during gamete formation. This leads to four possible combinations of alleles in the gametes (YR, Yr, yR, and yr), which explains the 9:3:3:1 phenotypic ratio observed in the F2 generation.

d. Summary of the Law of Independent Assortment

Alleles for different genes assort independently of one another during gamete formation.

The inheritance of one trait does not influence the inheritance of another, as long as the genes for those traits are on different chromosomes or far apart on the same chromosome.

3. Exceptions to Mendel's Laws

While Mendel's laws of inheritance explain many genetic patterns, there are several cases where inheritance does not follow these laws strictly:

a. Linked Genes

Genes that are located close together on the same chromosome are often inherited together because they do not assort independently. This is known as **genetic linkage**.

Linked genes violate the Law of Independent Assortment. However, **crossing over** during meiosis can sometimes separate linked genes, allowing them to assort independently.

b. Incomplete Dominance

In some cases, neither allele is completely dominant over the other, and the heterozygous phenotype is intermediate between the two homozygous phenotypes. This is known as **incomplete dominance**.

An example of incomplete dominance is seen in the snapdragon flower, where crossing red-flowered plants (RR) with white-flowered plants (rr) produces offspring with pink flowers (Rr).

c. Codominance

In **codominance**, both alleles in a heterozygous individual are fully expressed, resulting in a phenotype that shows both traits. An example is the **AB blood type** in humans, where both the A and B alleles are expressed.

d. Polygenic Inheritance

Some traits are controlled by **multiple genes**, each with a small additive effect on the phenotype. This is known as **polygenic inheritance** and often results in a continuous range of phenotypes, such as height, skin color, and eye color in humans.

e. Pleiotropy

In **pleiotropy**, a single gene affects multiple traits. For example, the gene responsible for **sickle cell anemia** affects not only red blood cell shape but also multiple systems in the body, leading to a variety of symptoms.

f. Epistasis

In some cases, one gene can interfere with or suppress the expression of another gene. This interaction between genes is known as **epistasis**. For example, in Labrador retrievers, a gene that determines coat color (black or brown) can be overridden by a second gene that controls whether any pigment is deposited at all (resulting in yellow fur).

4. Importance of Mendel's Laws

Mendel's Laws of Inheritance have far-reaching implications for understanding genetics and the transmission of traits across generations. These laws laid the groundwork for the study of heredity, genetic variation, and the role of alleles in determining an organism's phenotype. Mendel's work has also influenced other fields, such as **evolutionary biology**, **medicine**, and **agriculture**.

For instance:

In **medicine**, understanding the inheritance of genetic disorders like cystic fibrosis or sickle cell anemia has led to better diagnostic tools, genetic counseling, and potential treatments.

In **agriculture**, Mendel's principles have been applied to plant and animal breeding programs to improve crop yields, disease resistance, and other desirable traits.

Gregor Mendel's **Laws of Inheritance**—the **Law of Segregation** and the **Law of Independent Assortment**—form the foundation of classical genetics. These laws describe how alleles for traits are inherited from one generation to the next. While many traits follow Mendel's laws, modern genetics has revealed complexities, such as linked genes, polygenic traits, and epistasis, that extend beyond Mendelian inheritance. Nonetheless, Mendel's work remains a cornerstone of our understanding of biology and heredity.

Punnett Squares

A **Punnett square** is a simple diagram used in genetics to predict the genotypes and phenotypes of offspring resulting from a genetic cross. It was named after **Reginald C. Punnett**, an early 20th-century geneticist who developed the tool to better understand the transmission of traits based on **Mendelian inheritance**. Punnett squares are essential in predicting the outcome of genetic crosses by organizing parental alleles and showing how they combine in offspring.

Punnett squares can be used to analyze **monohybrid crosses** (one trait), **dihybrid crosses** (two traits), and even more complex cases involving multiple traits or different inheritance patterns, such as incomplete dominance, codominance, and sex-linked traits.

1. The Structure of a Punnett Square

A Punnett square is essentially a grid, with the parental gametes placed along the top and the side. The number of rows and columns in the Punnett square depends on the number of possible gametes each parent can produce. The gametes combine in the squares of the grid, representing all possible genotypes of the offspring.

Basic Structure of a Punnett Square for a Monohybrid Cross

In a monohybrid cross, each parent carries two alleles for a single trait, and these alleles segregate during gamete formation. The following is a simple example of a Punnett square used to determine the outcome of a cross involving one trait.

Rows and columns: One parent's gametes are listed on the top row, while the other parent's gametes are listed along the left column.

Intersection: Each box where a row and a column intersect represents the possible genotypes of the offspring, showing one allele from each parent.

For example, if we are crossing two heterozygous pea plants for seed color (Yy × Yy):

	Y (from Parent 1)	y (from Parent 1)
Y (from Parent 2)	YY	Yy
y (from Parent 2)	Yy	yy

The Punnett square shows that there is a **1/4** chance of having a homozygous dominant (YY), **1/2** chance of heterozygous (Yy), and **1/4** chance of homozygous recessive (yy) offspring.

This translates to a **3:1 ratio** of yellow-seeded (dominant phenotype) to green-seeded (recessive phenotype) offspring.

2. Monohybrid Cross: One Trait

A **monohybrid cross** is a genetic cross involving a single trait. Mendel's monohybrid crosses were designed to study the inheritance of one characteristic, such as **seed shape** or **seed color** in pea plants. Punnett squares are used to calculate the probable distribution of inherited traits in the offspring.

a. Example 1: Dominant and Recessive Alleles

Consider the cross between two pea plants that are heterozygous for seed color (Yy × Yy). Yellow (Y) is dominant, and green (y) is recessive.

	Y	y
Y	YY	Yy
y	Yy	yy

Genotypic ratio: 1 YY : 2 Yy : 1 yy

Phenotypic ratio: 3 Yellow : 1 Green

This simple 2x2 Punnett square shows that **75%** of the offspring will have yellow seeds (YY and Yy) and **25%** will have green seeds (yy).

b. Example 2: Homozygous Cross

If one parent is homozygous dominant (YY) and the other is homozygous recessive (yy), the cross (YY × yy) will produce only heterozygous (Yy) offspring.

Genotypic ratio: 100% Yy

Phenotypic ratio: 100% Yellow

All offspring will have the dominant yellow seed phenotype.

3. Dihybrid Cross: Two Traits

A **dihybrid cross** examines the inheritance of two different traits at the same time. Mendel's famous dihybrid cross involved seed shape and seed color, where round seeds (R) are dominant to wrinkled seeds (r), and yellow seeds (Y) are dominant to green seeds (y).

a. Example: Heterozygous Cross (YyRr × YyRr)

In this case, both parents are heterozygous for both traits (YyRr). To set up the Punnett square, we first determine the possible gametes each parent can produce. Each parent can produce the following four types of gametes: **YR, Yr, yR,** and **yr**. The resulting Punnett square will have 16 squares, as shown below:

	YR	Yr	yR	yr
YR	YYRR	YYRr	YyRR	YyRr
Yr	YYRr	YYrr	YyRr	Yyrr
yR	YyRR	YyRr	yyRR	yyRr
yr	YyRr	Yyrr	yyRr	yyrr

Phenotypic ratio: 9 Round Yellow : 3 Round Green : 3 Wrinkled Yellow : 1 Wrinkled Green

This 9:3:3:1 ratio is typical for a dihybrid cross involving two heterozygous parents and reflects the independent assortment of alleles during gamete formation, as per Mendel's **Law of Independent Assortment**.

4. Punnett Squares for Incomplete Dominance and Codominance

a. Incomplete Dominance

In **incomplete dominance**, neither allele is completely dominant over the other, and the heterozygous phenotype is intermediate between the two homozygous phenotypes. A common example involves snapdragon flowers, where the red allele (R) and the white allele (W) show incomplete dominance, resulting in pink flowers (RW).

	R	W
R	RR	RW
W	RW	WW

Genotypic ratio: 1 RR : 2 RW : 1 WW

Phenotypic ratio: 1 Red : 2 Pink : 1 White

In this case, crossing two pink-flowered snapdragons (RW × RW) produces offspring with **red**, **pink**, and **white** flowers.

b. Codominance

In **codominance**, both alleles are fully expressed in the heterozygous condition. A well-known example is the **AB blood type** in humans, where the A and B alleles are codominant.

	A	B
A	AA	AB
B	AB	BB

Genotypic ratio: 1 AA : 2 AB : 1 BB

Phenotypic ratio: 1 Type A : 2 Type AB : 1 Type B

In this case, individuals with **AB** blood type express both the A and B antigens on the surface of their red blood cells.

5. Punnett Squares for Sex-Linked Traits

Sex-linked traits are located on the **X chromosome**, and Punnett squares are especially useful in visualizing the inheritance of these traits. In humans, males have one X and one Y chromosome (XY), while females have two X chromosomes (XX).

a. Example: Color Blindness

Color blindness is an X-linked recessive trait. Let's examine a cross between a carrier female ($X^N X^c$, where X^N is normal vision and X^c is color blindness) and a normal male ($X^N Y$).

	X^N (from male)	Y (from male)
X^N (from female)	$X^N X^N$	$X^N Y$
X^c (from female)	$X^N X^c$	$X^c Y$

Phenotypic ratio: 1 Normal Female : 1 Carrier Female : 1 Normal Male : 1 Color-Blind Male

This shows that half the male offspring will be color-blind ($X^c Y$), while no female offspring will be color-blind, though some will be carriers ($X^N X^c$).

6. Using Punnett Squares in Predicting Genetic Outcomes

Punnett squares are valuable tools for predicting the possible genotypes and phenotypes of offspring from known parental genotypes. They allow geneticists, breeders, and researchers to estimate the probability of inheriting specific traits, including those controlled by simple Mendelian patterns, codominant traits, incomplete dominance, and sex-linked genes.

Punnett squares are fundamental tools in genetics, enabling the visualization of how alleles combine during reproduction to form different genotypes and phenotypes in offspring. By laying out the possible gametes and showing how they merge, Punnett squares allow the prediction of genetic outcomes in both simple and complex inheritance scenarios. From monohybrid to dihybrid crosses and traits like codominance or sex-linkage, Punnett squares continue to be essential in understanding the transmission of traits in living organisms.

Dominant and Recessive Traits

Dominant and recessive traits are foundational concepts in the field of genetics, first discovered by Gregor Mendel in the mid-19th century through his experiments on pea plants. These concepts explain how certain traits are passed down from parents to offspring and why some traits appear more frequently than others.

1. Mendel's Discoveries of Dominant and Recessive Traits

Gregor Mendel, known as the father of modern genetics, conducted a series of experiments in the 1860s to explore how traits are inherited from one generation to the next. Working with pea plants, Mendel focused on several specific traits, such as flower color, seed shape, and seed color. By cross-pollinating different varieties of peas, Mendel observed how these traits were passed on to their offspring.

One of his most critical discoveries was that traits can be **dominant** or **recessive**. He found that in some cases, one version of a trait (the dominant form) masked the presence of another version (the recessive form) in the offspring. These observations laid the foundation for what is now called **Mendelian inheritance**.

2. Definitions: Dominant and Recessive Traits

Dominant Trait: A dominant trait is one that is expressed or visible when an individual has at least one copy of the **dominant allele**. In a genetic pair, a dominant allele masks the effect of a recessive allele. It is represented by an uppercase letter (e.g., "A").

Recessive Trait: A recessive trait is one that is expressed only when both alleles are recessive, meaning the individual has two copies of the **recessive allele**. Recessive traits are masked by the presence of a dominant allele and are represented by a lowercase letter (e.g., "a").

For instance, in pea plants, purple flower color (P) is dominant over white flower color (p). A plant with the genotype PP or Pp will have purple flowers, while only plants with the genotype pp will have white flowers.

3. Alleles: The Basis for Dominant and Recessive Traits

Alleles are different versions of a gene. For each gene, an organism inherits two alleles, one from each parent. These alleles can either be the same or different.

Homozygous: When both alleles for a trait are identical (e.g., **PP** or **pp**), the organism is said to be **homozygous** for that trait.

Heterozygous: When the two alleles are different (e.g., **Pp**), the organism is **heterozygous**.

In heterozygous individuals, the dominant allele determines the organism's phenotype (the visible trait), while the recessive allele remains hidden or unexpressed.

4. Phenotype vs. Genotype

Genotype: The genetic makeup of an organism, which includes both dominant and recessive alleles. It refers to the combination of alleles an organism carries for a particular trait (e.g., PP, Pp, or pp for flower color).

Phenotype: The physical appearance or expression of a trait. The phenotype is determined by the genotype. For example, in pea plants, the genotype PP or Pp will result in a purple-flowered phenotype, while pp will result in a white-flowered phenotype.

In this case, the genotype determines whether a dominant or recessive trait is expressed in the phenotype.

5. Mendel's Experiments: Dominant and Recessive Traits in Pea Plants

Mendel used pea plants to observe how different traits were inherited. One of his most famous experiments involved crossing pea plants with different flower colors. By carefully controlling the pollination process, he discovered that certain traits, such as purple flower color, were consistently expressed, even when plants with white flowers were involved in the cross.

a. Mendel's Monohybrid Crosses

Mendel's **monohybrid crosses** involved plants that differed in only one trait. For example, he crossed a plant with purple flowers (PP) with a plant with white flowers (pp). In the first generation (F1), all the offspring had purple flowers (Pp), showing that the purple trait was **dominant**.

When Mendel allowed the F1 generation to self-pollinate, the resulting second generation (F2) had a **3:1 ratio** of purple to white flowers. This ratio indicated that the white flower trait (pp) was **recessive**, but it had reappeared in the F2 generation because the recessive allele had been passed down from both parents.

This discovery led Mendel to propose the concept of **dominance and recessiveness**, which forms the basis of modern genetics.

b. Law of Segregation

Mendel's **Law of Segregation** explains how alleles for a trait separate during the formation of gametes (sperm and egg cells). Each parent contributes one allele to their offspring. If the offspring inherits two dominant alleles or one dominant and one recessive allele, the dominant trait will be expressed. If the offspring inherits two recessive alleles, the recessive trait will be expressed.

6. Examples of Dominant and Recessive Traits in Humans

Many human traits follow the pattern of dominance and recessiveness, though inheritance can be more complex due to the involvement of multiple genes and environmental factors.

a. Dominant Traits in Humans

Widow's Peak: A pointed hairline at the center of the forehead is a dominant trait. If a person has one allele for a widow's peak, the trait will be expressed.

Freckles: Freckles are another dominant trait. A person only needs one dominant allele to have freckles.

Tongue Rolling: The ability to roll one's tongue into a U-shape is dominant. Individuals with at least one allele for tongue rolling can perform this action.

Dimples: Dimples in the cheeks are a dominant trait. People with one or two dominant alleles will have dimples.

b. Recessive Traits in Humans

Attached Earlobes: The trait for attached earlobes is recessive. A person needs two copies of the recessive allele to have attached earlobes.

Albinism: This genetic condition results from a lack of pigment in the skin, hair, and eyes. It is caused by recessive alleles.

Cystic Fibrosis: This is a recessive genetic disorder where a person must inherit two copies of the defective allele to develop the condition.

Sickle Cell Anemia: Another recessive disorder, sickle cell anemia occurs when an individual inherits two copies of the sickle cell allele, leading to abnormal red blood cells.

7. Incomplete Dominance and Codominance

While many traits follow the simple dominant-recessive pattern described by Mendel, there are exceptions. Some traits are governed by more complex inheritance patterns, such as **incomplete dominance** and **codominance**.

a. Incomplete Dominance

In **incomplete dominance**, neither allele is fully dominant over the other. As a result, the heterozygous genotype results in a blend of the two traits. For example, in snapdragon flowers, crossing a red flower (RR) with a white flower (WW) produces pink flowers (RW) in the offspring.

b. Codominance

In **codominance**, both alleles are equally expressed in the heterozygous condition. An example of codominance in humans is the **AB blood type**, where both the A and B alleles are expressed, leading to the presence of both A and B antigens on red blood cells.

8. Complexities in Inheritance Patterns

While Mendel's work laid the groundwork for understanding dominant and recessive traits, real-world genetics is often more complicated. For instance, some traits are influenced by multiple genes (**polygenic inheritance**), while others are affected by environmental factors. Additionally, some genes exhibit **pleiotropy**, where a single gene influences multiple traits, and **epistasis**, where one gene masks or modifies the expression of another gene.

9. Importance of Understanding Dominant and Recessive Traits

Understanding dominant and recessive traits is critical for predicting the inheritance of genetic conditions and traits in both plants and animals. It has applications in fields such as:

Medicine: Knowledge of how genetic diseases are inherited allows doctors to advise patients about the likelihood of passing on a disorder to their offspring.

Agriculture: Plant and animal breeders use knowledge of dominant and recessive traits to selectively breed organisms with desirable characteristics.

Genetic Counseling: Counselors use this information to help families understand their risk of inheriting or passing on genetic disorders.

Dominant and recessive traits form the basis of classical genetics. Through Mendel's experiments and the subsequent understanding of allele interactions, scientists have been able to predict the inheritance patterns of traits across generations. Dominant alleles mask recessive ones in heterozygous individuals, and Punnett squares can help visualize these inheritance patterns. While many traits follow simple Mendelian inheritance, others are governed by more complex mechanisms such as incomplete dominance and codominance, highlighting the richness of genetic diversity.

Chapter 9: Modern Genetics
Genetic Mutations

Genetic mutations are fundamental changes in the DNA sequence of an organism's genome. These alterations can impact the function of genes and subsequently influence an organism's phenotype. Understanding mutations is crucial for modern genetics, as they play a significant role in evolution, disease, and biotechnology.

1. Definitions and Types of Genetic Mutations

A **genetic mutation** is any change in the nucleotide sequence of DNA. Mutations can occur spontaneously due to errors in DNA replication or be induced by external factors such as chemicals or radiation. The effects of mutations on an organism can vary widely, from no noticeable impact to significant changes in phenotype or health.

Types of Mutations:

Point Mutations: These involve a change in a single nucleotide base pair.

Substitution: One nucleotide is replaced by another. This can be classified into:

Silent Mutation: A change that does not alter the amino acid sequence of the protein due to the redundancy in the genetic code. For example, changing a codon from GAA to GAG still codes for glutamic acid.

Missense Mutation: A change that results in the substitution of one amino acid for another in the protein. For example, changing a codon from GAG to GTG results in a different amino acid, potentially altering protein function.

Nonsense Mutation: A change that creates a premature stop codon, truncating the protein and often resulting in a nonfunctional protein. For example, changing a codon from CAG (glutamine) to UAG (stop) results in early termination of translation.

Insertions and Deletions (Indels): These mutations involve the addition or loss of one or more nucleotide base pairs.

Frameshift Mutation: Insertions or deletions that are not in multiples of three nucleotides shift the reading frame of the codons in the mRNA. This usually results in a completely different sequence of amino acids downstream from the mutation and often leads to a nonfunctional protein.

In-frame Mutations: Insertions or deletions that occur in multiples of three nucleotides do not shift the reading frame but can still affect protein function by adding or removing amino acids.

Duplication: A segment of DNA is copied and inserted into the genome, potentially leading to gene dosage effects or new gene functions.

Inversion: A segment of DNA is reversed end to end. This can disrupt gene function if the inversion occurs within a gene or affect gene expression if it alters regulatory regions.

Translocation: A segment of DNA is moved from one location to another within the genome or to a different chromosome. This can disrupt gene function or create fusion genes with altered or new functions.

2. Causes of Genetic Mutations

Mutations can arise from various sources:

Spontaneous Mutations: These occur without external influence, often due to errors in DNA replication or natural chemical changes in the DNA. For instance, spontaneous deamination of cytosine can result in a C-to-T transition mutation.

Induced Mutations: These result from exposure to environmental factors, such as:

Chemical Mutagens: Substances that can alter DNA structure, such as base analogs, alkylating agents, and intercalating agents.

Radiation: Ionizing radiation (e.g., X-rays, gamma rays) can cause breaks in DNA strands, while non-ionizing radiation (e.g., UV light) can lead to the formation of thymine dimers, which distort the DNA structure.

Biological Agents: Certain viruses can integrate their DNA into the host genome, potentially causing mutations.

3. Effects of Genetic Mutations

The impact of mutations can vary depending on their nature and location:

Neutral Mutations: Many mutations have no effect on the organism's fitness or phenotype. These are often silent mutations that do not change the protein's function.

Beneficial Mutations: Occasionally, a mutation can provide an advantage under certain environmental conditions, contributing to evolutionary processes. For example, mutations that confer antibiotic resistance in bacteria are beneficial in the presence of antibiotics.

Harmful Mutations: Mutations can lead to diseases or disorders if they disrupt essential biological functions. Some examples include:

Single-Gene Disorders: Such as cystic fibrosis (caused by mutations in the CFTR gene) and sickle cell anemia (caused by a mutation in the HBB gene).

Cancer: Many cancers are associated with mutations in genes that regulate cell growth and division. For example, mutations in the BRCA1 and BRCA2 genes increase the risk of breast and ovarian cancer.

4. Detection and Analysis of Mutations

Detecting and analyzing mutations is crucial for understanding genetic diseases, evolutionary biology, and biotechnology applications. Various methods are used:

Molecular Techniques:

Polymerase Chain Reaction (PCR): Amplifies specific DNA segments to detect mutations.

Sequencing: Determines the nucleotide sequence of a DNA segment to identify mutations. Techniques include Sanger sequencing and next-generation sequencing (NGS).

Genetic Screening: Used to identify individuals carrying mutations associated with genetic disorders, often before symptoms appear.

Genetic Testing: Includes tests for known mutations associated with specific conditions, such as BRCA testing for breast cancer susceptibility.

5. Genetic Mutations in Evolution

Mutations are a primary source of genetic variation within populations, which is essential for evolution. Beneficial mutations can provide adaptive advantages, leading to natural selection and the evolution of new traits or species. Understanding mutations helps elucidate how organisms adapt to their environments and how genetic diversity arises within populations.

6. Applications of Understanding Genetic Mutations

Medical Research and Treatment: Knowledge of genetic mutations contributes to developing targeted therapies, gene editing (e.g., CRISPR-Cas9), and personalized medicine approaches based on an individual's genetic profile.

Agriculture: Genetic engineering and selective breeding use understanding of mutations to develop crops with desirable traits, such as disease resistance or improved yield.

Forensic Science: DNA analysis, which includes detecting mutations, is used in criminal investigations and paternity testing.

Evolutionary Biology: Studying genetic mutations provides insights into the evolutionary history and relationships among species.

Genetic mutations are fundamental to genetics, influencing evolution, health, and biotechnology. By understanding the different types of mutations, their causes, effects, and methods for detection, scientists can better understand biological processes, develop new treatments, and advance various fields of research. The study of mutations remains a cornerstone of modern genetics, impacting everything from basic research to applied sciences.

Genetic Engineering and Biotechnology

Genetic engineering and **biotechnology** are transformative fields that apply principles of molecular biology and genetics to manipulate and harness the genetic material of organisms. These technologies have revolutionized medicine, agriculture, and industrial processes, and continue to drive significant advances in science and technology.

1. Genetic Engineering

Genetic engineering involves the direct manipulation of an organism's genome using biotechnology. This can include adding, removing, or altering genetic material in a way that does not occur naturally through mating or natural recombination. The goals of genetic engineering are diverse, including improving crop yields, developing new medical treatments, and producing industrial products.

a. Key Techniques in Genetic Engineering

Recombinant DNA Technology: This technique involves combining DNA from two different sources. A common method is to insert a gene of interest into a plasmid (a circular DNA molecule) and then introduce the plasmid into a host cell. This process allows for the production of proteins or other products by the host cell.

Restriction Enzymes: These are proteins that cut DNA at specific sequences, allowing for the precise insertion or removal of DNA segments.

Ligases: These enzymes join DNA fragments together, facilitating the creation of recombinant DNA molecules.

Gene Cloning: This process involves creating copies of a specific gene. A gene of interest is inserted into a vector (such as a plasmid) and introduced into a host cell. The host cell replicates, producing many copies of the gene. This is often used to produce large quantities of a protein.

CRISPR-Cas9: A revolutionary tool for genome editing that allows for precise modifications of DNA. The CRISPR-Cas9 system uses a guide RNA to target specific DNA sequences and the Cas9 protein to create double-strand breaks. This facilitates the addition or deletion of genetic material.

Gene Therapy: This technique aims to treat or prevent disease by introducing, removing, or altering genetic material within a patient's cells. Gene therapy can be used to correct genetic disorders, such as cystic fibrosis or muscular dystrophy, by delivering a functional copy of a gene or repairing a defective one.

Transgenic Organisms: These are organisms that have had a foreign gene inserted into their genome. For example, genetically modified (GM) crops often include genes for pest resistance or herbicide tolerance.

b. Applications of Genetic Engineering

Medicine:

Pharmaceutical Production: Genetic engineering enables the production of therapeutic proteins, such as insulin and growth hormones, through recombinant DNA technology.

Vaccines: Genetic engineering has been used to develop vaccines, including the production of proteins for use in vaccine formulations.

Agriculture:

Genetically Modified Crops: Crops can be engineered to be resistant to pests, diseases, or herbicides, or to improve nutritional content. Examples include Bt corn and Golden Rice.

Livestock: Genetic engineering is used to improve traits in livestock, such as disease resistance or faster growth rates.

Industrial Biotechnology:

Bioprocessing: Enzymes and microorganisms engineered to produce chemicals, fuels, or materials in an industrial setting.

Bioremediation: Using genetically modified organisms to clean up environmental pollutants.

2. Biotechnology

Biotechnology is a broader field that encompasses the use of biological systems and organisms to develop products and technologies. It includes genetic engineering but also involves other techniques that leverage biological processes.

a. Key Techniques in Biotechnology

Protein Engineering: The design and construction of new proteins with desirable properties. This can involve modifying the amino acid sequence of a protein to enhance its stability, activity, or specificity.

Synthetic Biology: An interdisciplinary field that combines biology and engineering to design and construct new biological parts, devices, and systems. It aims to create new biological functions or systems that do not exist in nature.

Metabolic Engineering: The optimization of metabolic pathways within microorganisms or plants to produce valuable chemicals, fuels, or materials. This involves modifying the organism's metabolism to enhance the production of specific compounds.

Bioprocessing: The use of living cells or their components to produce products. This includes fermentation processes used in the production of antibiotics, alcohol, and other chemicals.

b. Applications of Biotechnology

Medicine:

Diagnostic Tools: Biotechnology has developed advanced diagnostic tools, such as PCR (polymerase chain reaction) for detecting specific DNA sequences and biomarkers for disease diagnosis.

Personalized Medicine: Using genetic information to tailor treatments to individual patients, improving the efficacy and reducing side effects.

Agriculture:

Crop Improvement: Techniques like marker-assisted selection and genomics help improve crop varieties for better yield, disease resistance, and climate adaptability.

Biopesticides: The development of natural pest control methods, such as using beneficial microbes to target specific pests.

Environmental Biotechnology:

Waste Treatment: Using biotechnological processes to treat and manage waste, including the use of microorganisms to degrade pollutants.

Sustainable Practices: Developing biofuels and biodegradable materials to reduce reliance on fossil fuels and decrease environmental impact.

Food Industry:

Food Safety: Biotechnology is used to develop methods for detecting pathogens and ensuring the safety of food products.

Food Production: Techniques such as fermentation improve the production of various food products, including cheese, yogurt, and bread.

3. Ethical and Societal Considerations

The advancement of genetic engineering and biotechnology brings with it important ethical and societal considerations:

Genetic Privacy: With the ability to analyze and manipulate genetic information, concerns arise about privacy and the potential misuse of genetic data.

Biosafety: Ensuring that genetically modified organisms and biotechnological processes do not harm human health or the environment is crucial.

Ethical Use of Genetic Technologies: Debates continue about the appropriate use of genetic engineering, especially in areas like human germline editing and designer babies.

Equity and Access: Ensuring that the benefits of biotechnology and genetic engineering are accessible to all and do not exacerbate existing inequalities.

4. Future Directions in Genetic Engineering and Biotechnology

Gene Editing Advancements: Ongoing improvements to CRISPR and other gene-editing technologies aim to increase precision, reduce off-target effects, and explore new therapeutic applications.

Synthetic Genomics: Advances in synthetic biology and genome synthesis are paving the way for the creation of entirely new organisms with custom-designed genomes.

Regenerative Medicine: Combining genetic engineering with stem cell research to develop therapies for regenerative and repair purposes, including tissue and organ replacement.

Environmental Sustainability: Developing biotechnological solutions to address environmental challenges, such as climate change and resource depletion, through sustainable practices and innovative technologies.

Genetic engineering and biotechnology represent two of the most dynamic and impactful areas of modern science. By manipulating genetic material and harnessing biological systems, these fields have revolutionized medicine, agriculture, and industry. While they offer immense potential for improving human health, enhancing food security, and addressing environmental challenges, they also raise important ethical and societal questions. The ongoing development of these technologies promises continued advancements and new applications, underscoring the need for careful consideration of their implications and responsible use.

Ethical Considerations in Genetics

Ethical considerations in genetics are crucial as genetic research and technologies advance rapidly, impacting many areas of human life, from medicine and agriculture to criminal justice and personal privacy. Addressing these ethical issues helps ensure that genetic advancements benefit society while minimizing potential harms.

1. Ethical Issues in Genetic Research and Technology

Privacy and Confidentiality

Genetic Privacy: With the ability to sequence and analyze an individual's genome, there is significant concern about how genetic information is stored, shared, and used. There is a risk of genetic data being accessed without consent, leading to potential misuse by employers, insurance companies, or even unauthorized third parties.

Confidentiality: Ensuring that genetic information is kept confidential is essential. Breaches of confidentiality can lead to stigmatization or discrimination based on genetic predispositions.

Informed Consent

Complexity of Information: Genetic research often involves complex information about potential risks, benefits, and implications. Ensuring that participants fully understand this information and provide informed consent is crucial for ethical research practices.

Re-consent: As research progresses and new findings emerge, participants may need to be re-consented if new information significantly alters their understanding of the study's risks or benefits.

Genetic Discrimination

Insurance and Employment: There is a risk that genetic information could be used to discriminate against individuals in employment or insurance contexts. For example, individuals with a genetic predisposition to certain diseases might face higher insurance premiums or employment challenges.

Legislative Protections: Laws such as the Genetic Information Nondiscrimination Act (GINA) in the United States have been enacted to protect individuals from discrimination based on genetic information, though not all countries have similar protections.

2. Ethical Considerations in Genetic Testing and Screening

Predictive Testing

Impact of Results: Predictive genetic testing can reveal an individual's risk for developing certain genetic conditions, even if they are not currently symptomatic. This information can have psychological impacts and affect life planning.

Family Implications: Genetic testing results may also have implications for family members, as many genetic conditions are hereditary. Ensuring that individuals understand the implications for themselves and their relatives is important.

Prenatal and Newborn Screening

Ethical Dilemmas: Prenatal screening allows for the detection of genetic conditions in a fetus, leading to ethical questions about the use of this information. Decisions about continuation of pregnancy, based on genetic results, raise moral and ethical concerns.

Informed Decision-Making: Parents need accurate, unbiased information to make informed decisions about prenatal and newborn screening. This includes understanding the implications of test results and the potential options available.

3. Gene Editing and Therapy

Germline Editing

Heritable Changes: Gene editing technologies like CRISPR-Cas9 enable modifications to the germline (sperm, eggs, or embryos) that can be passed to future generations. This raises ethical concerns about unintended consequences and long-term impacts on the human gene pool.

Designer Babies: The possibility of creating "designer babies" with selected traits, such as enhanced intelligence or physical appearance, poses ethical questions about the limits of genetic modification and the potential for social inequality.

Somatic Gene Therapy

Therapeutic Goals: Somatic gene therapy involves modifying genes in somatic (non-reproductive) cells to treat or cure diseases. While this holds promise for treating genetic disorders, ethical considerations include ensuring that treatments are safe and effective.

Access and Equity: There are concerns about equitable access to gene therapies, as these treatments can be costly. Ensuring that advancements in gene therapy are accessible to all individuals, not just those with financial means, is a significant ethical issue.

4. Genetic Modification of Organisms

Genetically Modified Crops and Animals

Food Safety and Environmental Impact: The genetic modification of crops and animals for agriculture raises questions about food safety, environmental impact, and ecological balance. Ensuring rigorous testing and monitoring of genetically modified organisms (GMOs) is essential.

Labeling and Transparency: Ethical considerations include providing consumers with clear information about genetically modified products and ensuring transparency in labeling.

Ethical Treatment of Animals

Animal Welfare: Genetic modification of animals, including for research or agricultural purposes, raises concerns about animal welfare and ethical treatment. Ensuring humane practices and minimizing animal suffering is crucial.

5. Genetic Data and Human Rights

Ownership and Control

Data Ownership: Who owns genetic data—the individual from whom the data was obtained, the researchers who conducted the study, or the institution that funded the research? The ethical management of genetic data includes addressing issues of ownership and control.

Commercialization: The commercialization of genetic data, such as using it for developing products or patents, raises questions about the ethical use of personal genetic information and the potential for exploitation.

Global Health Equity

Access to Genetic Technologies: There is a need to address disparities in access to genetic technologies between different regions and populations. Ensuring that advancements in genetics benefit people globally, not just those in wealthy or developed nations, is a key ethical concern.

Benefit Sharing: Ethical considerations include how benefits from genetic research are shared with communities, especially those that contribute genetic information or participate in studies.

6. Ethical Frameworks and Guidelines

Ethical Review Boards

Institutional Review Boards (IRBs): Ethical review boards assess research proposals to ensure that they meet ethical standards, including obtaining informed consent, ensuring privacy, and minimizing risks to participants.

Ethical Guidelines: Various organizations and institutions have developed guidelines and codes of conduct for genetic research and applications. These include the Declaration of Helsinki, the Belmont Report, and guidelines from professional societies such as the American Society of Human Genetics.

Public Engagement

Public Dialogue: Engaging the public in discussions about genetic technologies and their implications is essential for ethical decision-making. Informed public debate helps address societal concerns and ensures that advancements in genetics align with societal values.

Education and Awareness: Promoting education and awareness about genetic technologies and their ethical implications can empower individuals to make informed decisions and participate in discussions about their use.

Ethical considerations in genetics are essential for guiding the responsible development and application of genetic technologies. As genetic research and biotechnology advance, addressing issues related to privacy, informed consent, discrimination, and the ethical use of genetic information is crucial. Ensuring that genetic advancements benefit society while minimizing risks and respecting individual rights requires ongoing dialogue, thoughtful consideration, and adherence to ethical guidelines. Balancing innovation with ethical responsibility will help ensure that genetic technologies are used in ways that promote human well-being and equity.

Part IV: Evolution and Diversity

Chapter 10: Principles of Evolution

Theories of Evolution

Theories of evolution explain the processes by which living organisms have changed over time and the mechanisms driving these changes. Several key theories have emerged, each contributing to our understanding of evolutionary biology. Here's an extensive discussion of the major theories and concepts related to evolution:

1. Historical Overview

Pre-Darwinian Views

Before Charles Darwin, several ideas about the origin and diversity of life existed, though they were often speculative and lacked empirical support. Notable pre-Darwinian views include:

Lamarckism: Proposed by Jean-Baptiste Lamarck in the early 19th century, Lamarckism suggested that organisms evolve through the inheritance of acquired characteristics. Lamarck theorized that traits developed during an organism's lifetime (e.g., a giraffe stretching its neck to reach higher leaves) could be passed on to offspring.

Catastrophism: Proposed by Georges Cuvier, this theory posited that the Earth has experienced a series of catastrophic events, such as floods and volcanic eruptions, leading to mass extinctions. Cuvier's ideas emphasized the role of catastrophic events in shaping the history of life on Earth.

2. Charles Darwin and Natural Selection

Darwin's Contributions

Charles Darwin, in his seminal work "On the Origin of Species" (1859), introduced the theory of **natural selection** as the primary mechanism of evolution. Darwin's theory was based on several key observations:

Variation: Darwin observed that individuals within a species exhibit variations in traits, such as size, color, and behavior. These variations can be inherited by offspring.

Overproduction: Most species produce more offspring than can survive to adulthood. This results in competition for resources.

Differential Survival and Reproduction: Individuals with advantageous traits are more likely to survive and reproduce, passing these traits on to the next generation. Over time, these advantageous traits become more common in the population.

Descent with Modification: Darwin proposed that species evolve over time through a process of gradual change. New species arise from common ancestors through modifications accumulated over long periods.

Evidence Supporting Natural Selection

Darwin's theory was supported by various lines of evidence, including:

Fossil Record: The fossil record provides evidence of gradual changes in species over time, showing transitional forms that bridge gaps between different groups of organisms.

Biogeography: The geographic distribution of species supports the idea of common descent. For example, Darwin's observations of finches on the Galápagos Islands demonstrated how isolated populations adapt to different environments.

Comparative Anatomy: The study of anatomical structures across different species reveals homologous structures (similar structures derived from a common ancestor) and vestigial organs (reduced structures with no current function).

Embryology: The similarities in embryonic development across different species provide evidence for common ancestry. For example, vertebrate embryos exhibit similar developmental stages early in development.

3. Modern Synthesis

The Modern Synthesis

The Modern Synthesis, developed in the early 20th century, integrates Darwinian natural selection with Mendelian genetics. Key contributors to the Modern Synthesis include:

Ronald A. Fisher: Fisher applied statistical methods to evolutionary biology, demonstrating how Mendelian genetics and natural selection can be combined to explain evolutionary change.

J.B.S. Haldane: Haldane's work on population genetics helped clarify how genetic variation is maintained and how selection acts on it.

Sewall Wright: Wright introduced concepts of genetic drift and effective population size, highlighting the role of random genetic changes in evolution.

Key Concepts of the Modern Synthesis

Population Genetics: The Modern Synthesis emphasizes the role of genetic variation within populations. Evolutionary change occurs through shifts in allele frequencies over time, driven by selection, genetic drift, gene flow, and mutation.

Genetic Drift: Random changes in allele frequencies in small populations due to chance events can lead to significant evolutionary changes. This is particularly important in small populations or in cases of population bottlenecks.

Gene Flow: The movement of genes between populations through migration and interbreeding can introduce new genetic material and affect evolutionary trajectories.

Mutation: Mutations are the source of new genetic variation. While most mutations are neutral or harmful, some can provide advantageous traits that may be selected for.

4. Post-Darwinian Theories and Developments

Neo-Darwinism

Neo-Darwinism, or the modern evolutionary synthesis, combines Darwinian natural selection with genetic principles. It emphasizes:

Genetic Variation: The role of genetic mutations and recombination in generating variation upon which natural selection acts.

Adaptive Evolution: The idea that adaptive traits arise through natural selection, leading to the evolution of traits that enhance an organism's fitness in a given environment.

Punctuated Equilibrium

Proposed by Stephen Jay Gould and Niles Eldredge in 1972, the theory of **punctuated equilibrium** challenges the notion of gradual evolution. It posits that:

Rapid Change: Evolution occurs in rapid bursts of change, often associated with speciation events, followed by long periods of stasis where species remain relatively unchanged.

Fossil Record: This theory suggests that the fossil record reflects these patterns, with gaps between sudden changes and long periods of little evolutionary change.

Evolutionary Developmental Biology (Evo-Devo)

Evo-Devo explores how developmental processes influence evolutionary change. Key concepts include:

Developmental Genes: The study of how changes in developmental genes and pathways lead to evolutionary changes in body plans and structures.

Heterochrony: The alteration in the timing or rate of developmental events can lead to significant morphological changes.

Modularity: The idea that evolutionary changes can occur in specific modules or parts of an organism, allowing for flexibility and innovation in evolution.

5. Sociobiology and Evolutionary Psychology

Sociobiology

Sociobiology, developed by E.O. Wilson, explores the evolutionary basis of social behaviors in animals, including humans. Key concepts include:

Altruism: Examining how altruistic behaviors (actions that benefit others at a cost to oneself) can evolve through kin selection, where individuals help relatives to increase the likelihood of shared genes being passed on.

Mate Selection: Investigating how evolutionary pressures shape mate selection and reproductive strategies.

Evolutionary Psychology

Evolutionary psychology applies evolutionary theory to understand human cognition and behavior. Key areas of focus include:

Adaptation: Investigating how psychological traits and behaviors may have evolved as adaptations to environmental pressures faced by our ancestors.

Universal Traits: Exploring common psychological traits across cultures and their evolutionary origins.

6. Criticisms and Alternatives

Criticisms of Evolutionary Theories

Intelligent Design: Critics argue that some features of organisms are too complex to have evolved through natural selection alone and propose an intelligent designer. This perspective is often considered outside the realm of scientific inquiry.

Creationism: Creationism rejects evolutionary theory in favor of a literal interpretation of religious texts regarding the origin of life and species.

Scientific Debate

The scientific community continues to debate and refine evolutionary theories based on new evidence. For example:

Epigenetics: The study of epigenetic changes (modifications to gene expression without altering DNA sequences) has led to discussions about how epigenetic factors might influence evolutionary processes.

Horizontal Gene Transfer: In microorganisms, horizontal gene transfer (the exchange of genetic material between organisms) challenges traditional views of evolution as a strictly vertical process.

The theories of evolution provide a framework for understanding the diversity of life on Earth and the mechanisms driving evolutionary change. From Darwin's theory of natural selection to modern concepts like punctuated equilibrium and evolutionary developmental biology, each theory contributes to a comprehensive understanding of how organisms evolve. Ongoing research and debate continue to refine our knowledge, highlighting the dynamic nature of evolutionary science. Understanding these theories not only sheds light on the history of life but also informs various fields, including medicine, conservation, and evolutionary psychology.

Natural Selection

Natural selection is a fundamental mechanism of evolution that explains how species adapt and evolve over time. First articulated by Charles Darwin in the mid-19th century, natural selection provides a framework for understanding how advantageous traits become more common in a population over generations. This concept has been refined and expanded with advances in genetics and evolutionary biology.

1. Basic Principles of Natural Selection

Variation

Genetic Variation: Within any population, individuals exhibit variations in traits due to genetic differences. These variations can include differences in physical characteristics, behaviors, and physiological processes.

Sources of Variation: Genetic variation arises from several sources:

Mutation: Random changes in DNA sequences can introduce new genetic variations.

Recombination: During sexual reproduction, genetic material is shuffled and recombined, creating new combinations of traits.

Gene Flow: Migration and interbreeding between populations can introduce new genes and variations.

Overproduction of Offspring

Exponential Growth: Most species produce more offspring than can survive to adulthood. This overproduction leads to competition for limited resources such as food, space, and mates.

Struggle for Existence: Individuals must compete for resources to survive and reproduce. This struggle is a key component of natural selection, as it creates a selective pressure on traits that enhance survival and reproductive success.

Differential Survival and Reproduction

Fitness: Fitness refers to an individual's ability to survive, reproduce, and pass on its genes to the next generation. Traits that improve an individual's fitness are considered advantageous.

Natural Selection in Action: Individuals with traits that enhance their fitness are more likely to survive and reproduce, passing those advantageous traits to their offspring. Conversely, individuals with less advantageous traits are less likely to survive and reproduce.

Adaptation

Adaptations: Over time, the advantageous traits become more common in the population, leading to evolutionary changes. These traits are called adaptations because they enhance an organism's ability to survive and reproduce in a specific environment.

Evolutionary Change: Adaptations result from the cumulative effect of natural selection acting on genetic variation. As environments change or new challenges arise, different traits may become advantageous, leading to ongoing evolutionary processes.

2. Mechanisms of Natural Selection

Stabilizing Selection

Definition: Stabilizing selection favors individuals with intermediate traits and selects against individuals with extreme traits. This type of selection tends to reduce variation and maintain the status quo.

Example: Human birth weight is often subject to stabilizing selection. Infants of very low or very high birth weight are at higher risk of health complications, so intermediate birth weights are more common.

Directional Selection

Definition: Directional selection favors individuals with traits at one extreme of the phenotype range. This type of selection shifts the average value of the trait in a specific direction.

Example: In a population of beetles, if larger beetles are better at escaping predators, directional selection may favor larger body sizes over time, leading to an increase in average beetle size.

Disruptive Selection

Definition: Disruptive selection favors individuals with extreme traits at both ends of the phenotype range and selects against individuals with intermediate traits. This can lead to increased variation and the potential for speciation.

Example: In environments with two distinct types of food resources, such as small and large seeds, birds with either very small or very large beaks may be favored over those with intermediate-sized beaks.

Sexual Selection

Definition: Sexual selection is a form of natural selection that arises from differences in reproductive success due to variation in traits related to mating success. It can lead to the evolution of traits that enhance an individual's attractiveness to mates or competitive ability.

Types of Sexual Selection:

Intersexual Selection: Often referred to as mate choice, where individuals of one sex (usually females) select mates based on specific traits, such as elaborate plumage in peacocks.

Intrasexual Selection: Competition between individuals of the same sex (usually males) for access to mates, such as antlers in deer used in combat.

Kin Selection

Definition: Kin selection is a form of natural selection where individuals increase their own genetic success by helping relatives. This is based on the idea that helping relatives can indirectly increase the likelihood of shared genes being passed on to the next generation.

Example: In some animal species, individuals will engage in altruistic behaviors, such as warning others of predators or helping to raise offspring, to benefit their relatives who share similar genes.

3. Evidence Supporting Natural Selection

Observational Evidence

Darwin's Finches: Darwin's observations of finches on the Galápagos Islands provided evidence for natural selection. The varying beak sizes and shapes of these finches were adapted to different food sources on the islands.

Peppered Moths: The change in coloration of peppered moths in response to industrial pollution is a classic example of natural selection. The darker form of the moth became more common in polluted areas where it had better camouflage.

Experimental Evidence

Laboratory Experiments: Experiments with microorganisms, such as bacteria and viruses, have demonstrated natural selection in action. For example, bacteria evolving resistance to antibiotics provide evidence of selection pressures and adaptation.

Artificial Selection: The selective breeding of plants and animals by humans to enhance desirable traits is a form of artificial selection that parallels natural selection. The diversity of domesticated breeds and varieties illustrates how selective pressures can drive evolutionary changes.

Fossil Record

Transitional Fossils: The fossil record provides evidence of gradual changes in species over time. Transitional fossils, such as those showing the evolution of whales from land-dwelling ancestors, support the idea of descent with modification through natural selection.

Stratigraphy: The layering of sediments and fossils in geological strata reveals the historical sequence of evolutionary changes and provides context for understanding how natural selection has shaped the diversity of life.

4. Modern Understanding and Extensions of Natural Selection

Modern Evolutionary Synthesis

Integration with Genetics: The Modern Synthesis integrates natural selection with principles of genetics, emphasizing that genetic variation and inheritance are central to evolutionary processes.

Population Genetics: The study of genetic variation within populations and how allele frequencies change over time due to natural selection, genetic drift, gene flow, and mutation.

Epigenetics

Epigenetic Variation: Recent research in epigenetics explores how changes in gene expression, without altering DNA sequences, can affect evolutionary processes. Epigenetic modifications can influence traits and may be heritable.

Evolutionary Developmental Biology (Evo-Devo)

Developmental Mechanisms: Evo-Devo studies how changes in developmental processes can lead to evolutionary changes in morphology and function. Understanding developmental pathways provides insights into how natural selection operates on developmental variations.

Human Evolution

Adaptations: Natural selection has shaped many aspects of human physiology and behavior. For example, adaptations related to skin pigmentation, disease resistance, and dietary preferences illustrate how natural selection has influenced human evolution.

Cultural Evolution: The concept of natural selection has also been extended to cultural evolution, where cultural traits and practices are subject to selective pressures and transmission through social learning.

Natural selection is a central mechanism of evolution that explains how advantageous traits increase in frequency within a population over time. Through the principles of variation, overproduction, differential survival and reproduction, and adaptation, natural selection drives the evolutionary process. Evidence from observational studies, experiments, and the fossil record supports the theory of natural selection, while modern extensions and integrations with genetics, epigenetics, and developmental biology continue to enhance our understanding of evolution. As a foundational concept in evolutionary biology, natural selection provides insights into the diversity of life and the processes shaping the natural world.

Speciation

Speciation is the evolutionary process by which populations evolve to become distinct species. It is a fundamental concept in evolutionary biology and explains the diversity of life forms observed on Earth. The process of speciation involves the splitting of a single evolutionary lineage into two or more genetically independent lineages. Understanding speciation provides insights into how biodiversity arises and how new species adapt to different environments.

1. Definitions and Basic Concepts

Species Concept

A **species** is commonly defined as a group of organisms that can interbreed and produce fertile offspring under natural conditions. However, the concept of a species can be complex and varies among different biological contexts. Several species concepts include:

Biological Species Concept: Defined by Ernst Mayr, it emphasizes reproductive isolation. According to this concept, species are groups of interbreeding natural populations that are reproductively isolated from other such groups.

Morphological Species Concept: Based on physical characteristics and structural features. This concept is useful for identifying species from fossil records or when reproductive data is not available.

Phylogenetic Species Concept: Defines species based on their evolutionary history and genetic distinctiveness. It uses phylogenetic trees to determine species boundaries based on shared derived traits.

Ecological Species Concept: Focuses on the ecological niche that a species occupies, emphasizing the role of ecological factors in defining species.

Mechanisms of Speciation

Speciation involves several mechanisms, which can be broadly categorized into two main types:

Allopatric Speciation: Occurs when populations are geographically separated by physical barriers, such as mountains or rivers, leading to genetic divergence. Over time, isolated populations adapt to their respective environments, accumulating genetic differences that can eventually lead to reproductive isolation.

Sympatric Speciation: Occurs without geographical separation, typically through mechanisms such as genetic mutations, polyploidy, or ecological specialization. In sympatric speciation, new species arise within the same geographic area due to differences in ecological niches or reproductive strategies.

2. Allopatric Speciation

Geographic Isolation

Barriers to Gene Flow: Physical barriers such as mountains, rivers, or oceanic islands can prevent gene flow between populations, leading to isolation. Geographic isolation reduces or eliminates interbreeding between populations, allowing them to evolve independently.

Founder Effect: When a small group of individuals establishes a new population in a different geographic area, the genetic diversity of the new population may be reduced compared to the original population. This founder effect can contribute to the divergence of the new population.

Adaptive Radiation

Rapid Diversification: Adaptive radiation is a form of allopatric speciation where a single ancestor species rapidly diversifies into a wide variety of forms to adapt to different ecological niches. This often occurs when a species colonizes a new environment with diverse habitats and resources.

Example: Darwin's finches on the Galápagos Islands are a classic example of adaptive radiation. Different finch species evolved distinct beak shapes and sizes to exploit various food sources on the islands.

Reproductive Isolation

Postzygotic Barriers: After mating occurs, reproductive isolation mechanisms may prevent hybrid offspring from developing into viable or fertile adults. These include hybrid inviability (offspring do not survive to adulthood) and hybrid sterility (offspring are sterile, as seen in mules).

Prezygotic Barriers: Before mating occurs, prezygotic barriers prevent fertilization. These include temporal isolation (different mating seasons), behavioral isolation (different mating rituals), and mechanical isolation (incompatible reproductive organs).

Sympatric Speciation

3Genetic Mutations

Chromosomal Mutations: Genetic mutations, including changes in chromosome number or structure, can lead to reproductive isolation. For example, polyploidy (having multiple sets of chromosomes) can create new species that cannot interbreed with their diploid ancestors.

Ecological Specialization: When individuals exploit different ecological niches within the same environment, they may develop distinct adaptations and preferences, leading to reproductive isolation and speciation.

Polyploidy

Definition: Polyploidy involves the multiplication of an organism's chromosome set, resulting in a genome with more than two sets of chromosomes. This is common in plants and can lead to speciation.

Autopolyploidy: Occurs when an individual has multiple sets of chromosomes derived from a single species. For example, in some plants, polyploid individuals can form new species with distinct traits.

Allopolyploidy: Involves hybridization between different species, followed by chromosome doubling. This can result in a new species with a genome that combines chromosomes from both parent species.

Disruptive Selection

Resource Partitioning: Disruptive selection can lead to sympatric speciation when individuals with extreme traits have higher fitness in different ecological niches. Over time, these divergent traits can lead to reproductive isolation.

Example: In cichlid fishes in African lakes, different species have evolved to exploit various food resources, leading to a high degree of ecological specialization and speciation within the same lake.

4. Hybridization and Speciation

Hybrid Zones

Definition: Hybrid zones are regions where the ranges of two or more species overlap, allowing for interbreeding and hybrid formation. These zones can provide insights into the dynamics of speciation and reproductive isolation.

Stable Hybrid Zones: In some cases, hybrids may persist and occupy specific ecological niches, contributing to the maintenance of hybrid zones.

Hybrid Speciation

New Species Formation: In some instances, hybrids between different species can themselves become new species if they are reproductively isolated from both parent species. This can occur through mechanisms such as polyploidy or ecological adaptation.

Example: The hybridization of certain plant species can lead to new polyploid species with unique traits and reproductive isolation from parent species.

5. Speciation and Evolutionary Theory

Evidence from Molecular Genetics

Genetic Divergence: Molecular studies, including DNA sequencing, provide evidence for speciation by revealing genetic differences between species. Phylogenetic analysis can trace the evolutionary relationships between species and identify patterns of divergence.

Genomic Data: Advances in genomics allow for the study of speciation at the molecular level, including identifying genes associated with reproductive isolation and adaptive traits.

Ecological and Environmental Factors

Environmental Changes: Changes in the environment, such as climate change or habitat fragmentation, can drive speciation by creating new ecological niches and altering selective pressures.

Human Impact: Human activities, including habitat destruction and introduction of invasive species, can influence speciation processes by altering ecological interactions and creating new selective pressures.

6. Current Research and Future Directions

Genomics and Speciation

Comparative Genomics: Researchers use comparative genomics to study the genetic basis of speciation and identify genetic changes associated with the emergence of new species.

Genomic Tools: Advances in genomic tools, such as CRISPR and high-throughput sequencing, allow for detailed studies of genetic variations and their roles in speciation.

Evolutionary Developmental Biology

Developmental Pathways: Evo-Devo research explores how changes in developmental pathways contribute to the evolution of new species. Understanding developmental mechanisms can provide insights into the origins of reproductive isolation and adaptive traits.

Conservation and Speciation

Conservation Efforts: Studying speciation processes can inform conservation strategies by understanding how populations adapt to changing environments and identifying conservation priorities for endangered species.

Speciation is a central concept in evolutionary biology that explains how new species arise and diversify. It involves mechanisms such as allopatric and sympatric speciation, genetic mutations, polyploidy, and hybridization. Understanding speciation provides insights into the diversity of life and the processes driving evolutionary change. Advances in genomics, evolutionary developmental biology, and conservation research continue to enhance our understanding of speciation and its implications for biodiversity and adaptation.

Chapter 11: Classification of Living Organisms

The Five Kingdoms

The classification of living organisms into five kingdoms represents a historical and foundational framework in biology for understanding the diversity of life on Earth. This system, which was first proposed in the 1960s by Robert Whittaker, categorizes life into five broad groups based on key characteristics such as cell structure, mode of nutrition, and reproductive methods. The five kingdoms are: Monera, Protista, Fungi, Plantae, and Animalia.

A. Kingdom Monera

Overview

Cell Type: Monera includes prokaryotic organisms, which lack a true nucleus and membrane-bound organelles. Their genetic material is not enclosed within a nuclear envelope but is instead found in a single circular chromosome located in the cytoplasm.

Diversity: This kingdom encompasses bacteria and archaea, two fundamentally different domains of life. Although they share some similarities, such as their prokaryotic cell structure, they differ significantly in their genetic, biochemical, and ecological characteristics.

Bacteria

Characteristics: Bacteria are unicellular organisms with a variety of shapes (cocci, bacilli, spirilla) and metabolic modes (autotrophic, heterotrophic). They can thrive in diverse environments, including extreme conditions.

Reproduction: Bacteria reproduce asexually through binary fission, where a single cell divides into two identical daughter cells.

Ecological Roles: Bacteria play crucial roles in nutrient cycling, such as nitrogen fixation, decomposition, and bioremediation.

Archaea

Characteristics: Archaea are similar in size and shape to bacteria but differ in their genetic sequences and biochemical pathways. They often inhabit extreme environments, such as hot springs, salt lakes, and acidic environments.

Metabolism: Archaea include methanogens (producing methane), halophiles (salt-loving), and thermophiles (heat-loving). Their metabolic processes can differ significantly from those of bacteria.

Evolutionary Significance: Archaea are thought to be more closely related to eukaryotes than to bacteria, based on genetic and molecular evidence.

B. Kingdom Protista

Overview

Cell Type: Protists are eukaryotic organisms with a true nucleus and membrane-bound organelles. This kingdom includes a diverse array of mostly unicellular organisms, though some are multicellular.

Diversity: Protists are classified into three main groups based on their modes of nutrition and movement: protozoa (animal-like), algae (plant-like), and slime molds (fungus-like).

Protozoa

Characteristics: Protozoa are primarily heterotrophic, obtaining nutrients by ingestion. They exhibit various forms of locomotion, including flagella, cilia, or pseudopodia.

Examples: Amoebas, paramecia, and giardia are examples of protozoa. Some protozoa are free-living, while others are parasitic.

Algae

Characteristics: Algae are primarily autotrophic, capable of photosynthesis. They can be unicellular or multicellular and are found in aquatic environments.

Types: Algae include phytoplankton, green algae, brown algae, and red algae. They are important primary producers in aquatic ecosystems.

Examples: Chlorella, Euglena, and giant kelp are examples of algae.

Slime Molds

Characteristics: Slime molds exhibit characteristics of both fungi and amoebas. They are primarily decomposers and can exist in two phases: a free-living amoeba-like phase and a multicellular spore-producing phase.

Examples: Plasmodial slime molds and cellular slime molds are examples, which play roles in nutrient cycling and decomposition.

C. Kingdom Fungi

Overview

Cell Type: Fungi are eukaryotic organisms with cell walls made of chitin. They are primarily non-motile and reproduce both sexually and asexually.

Diversity: This kingdom includes molds, yeasts, and mushrooms. Fungi are classified based on their reproductive structures and life cycles.

Structure and Function

Hyphae: Fungi consist of a network of filaments called hyphae, which form a mass known as mycelium. The hyphae absorb nutrients from the environment.

Reproduction: Fungi reproduce through spores, which can be produced either sexually (through the fusion of specialized sexual structures) or asexually (through budding or spore release).

Ecological Roles

Decomposers: Fungi play a vital role in decomposition, breaking down organic matter and recycling nutrients in ecosystems.

Symbiosis: Many fungi form symbiotic relationships with other organisms. For example, mycorrhizal fungi associate with plant roots to enhance nutrient uptake, and lichens are symbiotic partnerships between fungi and algae or cyanobacteria.

D. Kingdom Plantae

Overview

Cell Type: Plants are eukaryotic organisms with cell walls made of cellulose. They are predominantly autotrophic, performing photosynthesis to produce their own food.

Diversity: The plant kingdom is divided into several major groups, including bryophytes, pteridophytes, gymnosperms, and angiosperms.

Bryophytes

Characteristics: Bryophytes include mosses, liverworts, and hornworts. They are non-vascular plants and typically grow in moist environments.

Reproduction: Bryophytes reproduce through spores and have a dominant gametophyte generation in their life cycle.

Pteridophytes

Characteristics: Pteridophytes include ferns, horsetails, and club mosses. They are vascular plants with specialized tissues for transporting water and nutrients.

Reproduction: Pteridophytes reproduce via spores and have a dominant sporophyte generation in their life cycle.

Gymnosperms

Characteristics: Gymnosperms include conifers, cycads, ginkgoes, and gnetophytes. They produce seeds that are not enclosed within a fruit.

Reproduction: Gymnosperms reproduce through seeds and cones, with a dominant sporophyte generation.

Angiosperms

Characteristics: Angiosperms, or flowering plants, produce seeds enclosed within fruits. They exhibit a wide range of forms, from grasses to flowering trees.

Reproduction: Angiosperms reproduce through flowers and seeds, with a dominant sporophyte generation. They rely on various mechanisms, including pollinators and wind, for fertilization.

E. Kingdom Animalia

Overview

Cell Type: Animals are eukaryotic, multicellular organisms that lack cell walls. They are heterotrophic, obtaining nutrients by ingesting other organisms.

Diversity: The animal kingdom is the most diverse, encompassing a wide range of organisms from simple sponges to complex mammals. Animals are classified into various phyla based on body plans, symmetry, and developmental patterns.

Major Phyla

Porifera: Sponges are simple animals with a porous body structure and lack true tissues and organs. They filter feed by drawing water through their pores.

Cnidaria: This phylum includes jellyfish, corals, and sea anemones. Cnidarians have radial symmetry and specialized stinging cells called nematocysts.

Platyhelminthes: Flatworms, including planarians, tapeworms, and flukes, are characterized by their flattened bodies and bilateral symmetry. Some are parasitic.

Annelida: Annelids, or segmented worms, include earthworms, leeches, and polychaetes. They have segmented bodies and a coelom (body cavity).

Arthropoda: Arthropods are the largest animal phylum and include insects, arachnids (spiders, mites), and crustaceans (crabs, shrimp). They have an exoskeleton, jointed legs, and segmented bodies.

Chordata: Chordates include vertebrates (fish, amphibians, reptiles, birds, mammals) and some invertebrates (lancelets, tunicates). They are characterized by the presence of a notochord, a dorsal nerve cord, and pharyngeal slits.

Reproduction and Development

Reproduction: Most animals reproduce sexually, with variations in reproductive strategies, including internal and external fertilization, and a range of developmental patterns (e.g., direct development, metamorphosis).

Development: Animal development can be classified into two main types:

Protostome Development: In protostomes, the mouth develops first from the blastopore, and the fate of embryonic cells is determined early.

Deuterostome Development: In deuterostomes, the anus develops first from the blastopore, and the fate of embryonic cells is more flexible.

F. Modern Perspectives and Challenges

Genomics and Speciation

Genetic Evidence: Advances in molecular phylogenetics and DNA sequencing have led to a more nuanced understanding of relationships among kingdoms. The traditional five-kingdom system is being revised to reflect insights from molecular data.

Three Domains of Life: The concept of three domains (Bacteria, Archaea, Eukarya) is now widely accepted, which further refines our understanding of the diversity of life beyond the five-kingdom system.

The Role of Endosymbiosis

Endosymbiotic Theory: The theory of endosymbiosis explains the origin of eukaryotic cells, including the role of symbiotic relationships in the evolution of organelles like mitochondria and chloroplasts.

Implications: Understanding endosymbiosis has implications for the classification and evolutionary relationships among eukaryotic kingdoms.

Environmental and Ecological Considerations

Biodiversity Conservation: The classification system helps in understanding and conserving biodiversity, which is crucial for ecosystem health and resilience.

Impact of Climate Change: Climate change affects the distribution and behavior of organisms across different kingdoms, influencing conservation strategies and our understanding of species interactions.

The classification of living organisms into the five kingdoms represents a foundational framework for organizing the diversity of life. Each kingdom is characterized by distinct cellular structures, modes of nutrition, and reproductive strategies. While the traditional five-kingdom system provides a broad overview, modern advances in molecular biology and evolutionary research continue to refine and expand our understanding of life's diversity. The integration of new genetic and ecological data helps in providing a more accurate and dynamic view of the relationships among living organisms.

The Three Domains of Life

The classification of life into three domains is a modern framework that reflects our current understanding of the evolutionary relationships among living organisms. This system was introduced in 1990 by Carl Woese and his colleagues based on molecular and genetic evidence. The three domains of life are: Bacteria, Archaea, and Eukarya. Each domain represents a major branch of the tree of life and encompasses a wide range of organisms with distinct cellular and molecular characteristics.

A. Domain Bacteria

Overview

Cell Type: Bacteria are prokaryotic cells, meaning they lack a membrane-bound nucleus and organelles. Their genetic material is typically a single circular DNA molecule located in the cytoplasm, not enclosed in a nuclear envelope.

Structure: Bacterial cells have a simple structure compared to eukaryotic cells. They are surrounded by a plasma membrane and often have a rigid cell wall composed of peptidoglycan. Some bacteria have additional structures such as pili, flagella, and capsules.

Diversity

Morphology: Bacteria exhibit diverse shapes, including spherical (cocci), rod-shaped (bacilli), and spiral (spirilla) forms.

Metabolic Diversity: Bacteria display a wide range of metabolic types. They can be autotrophic (producing their own food through processes like photosynthesis or chemosynthesis) or heterotrophic (obtaining nutrients from other organisms). They can also be aerobic (requiring oxygen) or anaerobic (not requiring oxygen).

Habitat: Bacteria are found in nearly every environment on Earth, from extreme conditions (hot springs, deep-sea vents) to more common habitats (soil, water, human body).

Reproduction

Binary Fission: Bacteria reproduce asexually through binary fission, where a single cell divides into two identical daughter cells. This process is rapid and allows bacterial populations to grow quickly.

Significance

Ecological Roles: Bacteria play essential roles in nutrient cycling, such as nitrogen fixation in the soil, decomposition of organic matter, and bioremediation (cleaning up environmental pollutants).

Medical Importance: Some bacteria are pathogenic and cause diseases, while others are beneficial and used in biotechnology and medicine, such as probiotics and antibiotics production.

B. Domain Archaea

Overview

Cell Type: Archaea are also prokaryotic, like bacteria, but they have distinct molecular and biochemical features that set them apart from bacteria. They lack a nucleus and membrane-bound organelles.

Structure: Archaeal cell walls do not contain peptidoglycan but instead may have pseudopeptidoglycan or other unique polysaccharides. Archaea also have distinct lipid membranes, which include ether bonds rather than ester bonds found in bacterial and eukaryotic membranes.

Diversity

Extremophiles: Many archaea are extremophiles, thriving in extreme environments such as high-temperature hot springs (thermophiles), highly saline environments (halophiles), or acidic conditions (acidophiles).

Methanogens: A subgroup of archaea known as methanogens produce methane as a metabolic byproduct. They are commonly found in the guts of ruminants and in anaerobic environments such as swamps and marshes.

Non-extremophiles: Some archaea live in more moderate environments and can be found in soil and oceans.

Reproduction

Binary Fission: Like bacteria, archaea reproduce asexually through binary fission. However, their replication machinery and processes are more similar to eukaryotes than to bacteria.

Significance

Ecological Roles: Archaea play a crucial role in global biogeochemical cycles, such as methane production and carbon cycling. Their unique metabolic processes are significant in understanding extreme environments and potential applications in biotechnology.

Evolutionary Insights: Archaea are considered to be more closely related to eukaryotes than to bacteria, based on similarities in their genetic and biochemical features.

C. Domain Eukarya

Overview

Cell Type: Eukaryotic cells have a true nucleus enclosed by a nuclear membrane and contain membrane-bound organelles such as mitochondria, chloroplasts, and the endoplasmic reticulum.

Structure: Eukaryotic cells are generally more complex than prokaryotic cells. They have a cytoskeleton that provides structural support, and many have cell walls (e.g., plants and fungi) or extracellular matrices (e.g., animals).

Diversity

Major Kingdoms: The domain Eukarya is divided into several kingdoms:

Protista: A diverse group of mostly unicellular organisms with varied modes of nutrition (e.g., protozoa, algae, slime molds).

Fungi: Includes molds, yeasts, and mushrooms, characterized by chitin in their cell walls and heterotrophic nutrition.

Plantae: Multicellular, autotrophic organisms with cell walls made of cellulose, capable of photosynthesis (e.g., mosses, ferns, flowering plants).

Animalia: Multicellular, heterotrophic organisms without cell walls, typically with complex tissues and organs (e.g., sponges, insects, mammals).

Reproduction

Sexual and Asexual Reproduction: Eukaryotes can reproduce both sexually (through the fusion of gametes) and asexually (through processes such as mitosis, budding, or vegetative propagation).

Significance

Ecological Roles: Eukaryotes play diverse roles in ecosystems, from producers (plants) to decomposers (fungi) to consumers (animals). Their interactions shape ecosystems and influence biogeochemical cycles.

Medical and Biotechnological Applications: Eukaryotic cells, particularly those of plants, fungi, and animals, are crucial in medicine and biotechnology. They are sources of drugs, agricultural products, and industrial enzymes.

D. Comparative Analysis

1. Molecular Differences

Genetic Material: Bacteria and archaea have circular DNA without histones, while eukaryotes have linear DNA associated with histones.

Ribosomes: The ribosomes of bacteria and archaea are similar but differ from those of eukaryotes in size and sensitivity to antibiotics.

2. Cellular Processes

Metabolism: Eukaryotes often have more complex metabolic pathways compared to prokaryotes. For instance, mitochondria and chloroplasts in eukaryotes are specialized for energy production and photosynthesis, respectively.

Reproduction: Eukaryotic reproduction is more varied, with complex mechanisms for sexual reproduction and cell division compared to the simpler binary fission in prokaryotes.

E. Modern Perspectives

1. Evolutionary Relationships

Tree of Life: The three-domain system provides a framework for understanding the evolutionary relationships among all life forms, with a focus on genetic and molecular evidence.

2. Advances in Genomics

Sequencing Technologies: Advances in genomic sequencing and bioinformatics continue to refine our understanding of the domains and their relationships, leading to ongoing revisions in classification.

3. Implications for Research

Biotechnological Innovations: The study of extremophiles in Archaea and the diversity of Eukarya contributes to advances in biotechnology, medicine, and environmental science.

The classification of life into the three domains—Bacteria, Archaea, and Eukarya—provides a comprehensive framework for understanding the fundamental differences and similarities among organisms. This system, based on genetic and molecular evidence, reflects the evolutionary history and relationships of life forms. As research progresses, the classification continues to evolve, enhancing our knowledge of the diversity of life and its complexities.

Taxonomy and Phylogeny

Taxonomy and phylogeny are fundamental aspects of biology that deal with the classification, naming, and evolutionary relationships of organisms. They provide a structured framework for understanding the diversity of life and its evolutionary history. While taxonomy focuses on the classification and naming of organisms, phylogeny explores their evolutionary relationships and history. Here's an extensive discussion of both concepts:

A. Taxonomy

1. Definition and Purpose

Definition: Taxonomy is the science of naming, describing, and classifying organisms into hierarchical categories based on their characteristics, evolutionary history, and relationships. It aims to organize biological diversity into a structured system.

Purpose: The main goals of taxonomy are to provide a universal naming system, to categorize organisms into meaningful groups, and to facilitate the study and communication of biological information.

2. Taxonomic Ranks

Taxonomic classification is hierarchical, with each rank representing a different level of organization. The major taxonomic ranks from broadest to most specific are:

Domain: The highest and broadest rank, which divides life into three domains—Bacteria, Archaea, and Eukarya.

Kingdom: The second-highest rank, dividing organisms into major groups such as Monera, Protista, Fungi, Plantae, and Animalia (or revised groups in modern systems).

Phylum (or **Division** in plants): Groups organisms based on major body plans and structural features.

Class: Divides organisms within a phylum based on additional shared characteristics.

Order: Further divides classes into groups with more specific similarities.

Family: Groups organisms within an order that share even closer characteristics.

Genus: Groups species that are closely related and share a common ancestor.

Species: The most specific rank, defining a group of organisms that can interbreed and produce fertile offspring. It is the basic unit of classification and the fundamental category in taxonomy.

3. Taxonomic Hierarchy Example

Domain: Eukarya

Kingdom: Animalia

Phylum: Chordata

Class: Mammalia

Order: Carnivora

Family: Felidae

Genus: Panthera

Species: Panthera leo (lion)

4. Nomenclature

Binomial Nomenclature: Introduced by Carl Linnaeus, this system assigns each species a two-part scientific name consisting of the genus name and the species name (e.g., *Homo sapiens*). It provides a standardized and universally accepted naming convention.

Rules and Conventions: Scientific names are typically in Latin or Greek, and the rules for naming are governed by international codes (e.g., International Code of Botanical Nomenclature for plants, International Code of Zoological Nomenclature for animals).

5. Modern Taxonomy

Molecular Phylogenetics: Advances in DNA sequencing and molecular techniques have led to revisions in taxonomy. Genetic data often reveal new relationships and require adjustments to traditional classifications.

Cladistics: A method of classification based on common ancestry and evolutionary relationships. Cladistics focuses on identifying shared derived characteristics (synapomorphies) to define groups (clades).

B. Phylogeny

1. Definition and Purpose

Definition: Phylogeny is the study of the evolutionary history and relationships among organisms. It seeks to reconstruct the evolutionary tree (phylogenetic tree) that illustrates how different species are related through common ancestry.

Purpose: The aim of phylogenetics is to understand the evolutionary processes that have led to the diversity of life, to uncover the patterns of evolution, and to elucidate the historical relationships among organisms.

2. Phylogenetic Tree

Structure: A phylogenetic tree is a diagram representing evolutionary relationships among species. It consists of nodes (representing common ancestors) and branches (representing evolutionary lineages).

Cladogram: A type of phylogenetic tree that shows the relationships among species based on shared derived characteristics but does not represent evolutionary time.

Phylogram: A type of phylogenetic tree where branch lengths represent the amount of evolutionary change or time.

3. Methods of Phylogenetic Analysis

Morphological Data: Traditional methods of phylogenetic analysis relied on physical characteristics and anatomical features. However, this approach is limited by the subjective interpretation of traits.

Molecular Data: Modern phylogenetics often uses genetic, molecular, and biochemical data, such as DNA, RNA, and protein sequences, to construct more accurate and objective evolutionary relationships.

Comparative Genomics: The comparison of whole genomes or specific genes among different organisms helps to identify evolutionary relationships and trace the history of gene families and functions.

4. Phylogenetic Classification

Monophyletic Groups: Taxonomic groups that include a common ancestor and all of its descendants. Monophyletic groups are considered to accurately reflect evolutionary relationships.

Paraphyletic Groups: Taxonomic groups that include a common ancestor but not all of its descendants. Paraphyletic groups are less ideal because they do not fully represent evolutionary history.

Polyphyletic Groups: Taxonomic groups that do not include the most recent common ancestor of all members. Polyphyletic groups are based on convergent characteristics rather than shared ancestry.

5. Evolutionary Concepts

Common Ancestry: All organisms share a common ancestry, and phylogenetic trees illustrate how different lineages have diverged from shared ancestors over time.

Adaptive Radiation: The rapid diversification of a lineage into a wide variety of forms to adapt to different ecological niches. This process is often illustrated by phylogenetic trees showing the branching patterns of adaptive evolution.

Evolutionary Reversals: Occurrences where traits revert to a previous state, which can complicate phylogenetic analysis and require careful interpretation of evolutionary relationships.

6. Applications of Phylogeny

Understanding Biodiversity: Phylogenetic analysis helps to reveal the relationships among species, improving our understanding of biodiversity and the evolutionary processes that shape it.

Conservation Biology: Phylogenetic information is used to identify and prioritize species for conservation based on their evolutionary significance and the preservation of genetic diversity.

Medical Research: Phylogenetics contributes to understanding the evolution of pathogens, tracking the spread of diseases, and developing vaccines and treatments based on evolutionary insights.

C. Integration of Taxonomy and Phylogeny

1. Evolutionary Classification

Phylogenetic Classification: Modern taxonomy increasingly incorporates phylogenetic principles to create classifications that reflect evolutionary relationships. This approach integrates morphological, molecular, and genetic data to produce more accurate and meaningful classifications.

2. Dynamic Nature of Classification

Revisions and Updates: Taxonomy and phylogeny are dynamic fields, and classifications are subject to change as new data and technologies emerge. Taxonomic revisions based on phylogenetic analysis can lead to reclassification and renaming of species and groups.

3. Challenges and Future Directions

Incomplete Data: Phylogenetic analysis can be limited by incomplete or ambiguous data, requiring ongoing refinement and validation of evolutionary relationships.

Integration of Multiple Data Sources: Combining data from various sources, such as morphological traits, molecular sequences, and ecological information, provides a more comprehensive understanding of evolutionary relationships.

Global Efforts: Collaborative international efforts and large-scale projects, such as genome sequencing and biodiversity databases, continue to enhance our knowledge of taxonomy and phylogeny, driving advancements in biological research and classification.

Taxonomy and phylogeny are fundamental to the study of biology, providing a framework for classifying and understanding the diversity of life. Taxonomy organizes organisms into hierarchical categories based on their characteristics, while phylogeny reveals their evolutionary relationships and history. Advances in molecular biology and genetic analysis have revolutionized these fields, leading to more accurate and refined classifications. The integration of taxonomy and phylogeny continues to enhance our understanding of life's complexity and evolutionary processes.

Protists

Bacteria and **Archaea** are two of the three domains of life and represent some of the most ancient and diverse forms of life on Earth. While both are prokaryotic—meaning they lack a membrane-bound nucleus and organelles—they have distinct differences in their genetic, biochemical, and ecological characteristics. This chapter provides an extensive discussion on the nature, classification, and significance of Bacteria and Archaea.

Bacteria

1. Overview

Bacteria are prokaryotic organisms with a simple cell structure. They lack a true nucleus and membrane-bound organelles. The genetic material in bacteria consists of a single, circular DNA molecule located in the nucleoid region of the cell. Some bacteria may also contain plasmids—small, circular DNA molecules that can carry additional genes.

2. Structure

Most bacteria have a cell wall composed of peptidoglycan, a polymer consisting of sugars and amino acids. The cell wall provides structural support and protection. Gram-positive bacteria have a thick peptidoglycan layer that retains the Gram stain and appears purple under a microscope. Gram-negative bacteria have a thinner peptidoglycan layer and an additional outer membrane, which contains lipopolysaccharides (LPS) and does not retain the Gram stain, appearing pink.

The plasma membrane surrounds the cell wall and regulates the movement of substances in and out of the cell. It is composed of a phospholipid bilayer with embedded proteins. Many bacteria have flagella (long, whip-like structures) for movement, and pili (short, hair-like projections) for attachment to surfaces and other cells. Some bacteria have a capsule, a thick outer layer that protects them from desiccation and immune system attacks.

3. Classification

Bacteria can be classified based on their shape:

Cocci: Spherical bacteria (e.g., *Streptococcus*).

Bacilli: Rod-shaped bacteria (e.g., *Escherichia coli*).

Spirilla: Spiral-shaped bacteria (e.g., *Helicobacter pylori*).

Bacteria are also classified based on their metabolism:

Autotrophs: Obtain energy from inorganic sources, such as cyanobacteria that perform photosynthesis.

Heterotrophs: Obtain energy from organic compounds, including many bacteria that decompose organic matter.

Oxygen requirements also classify bacteria:

Aerobic: Require oxygen for growth (e.g., *Mycobacterium tuberculosis*).

Anaerobic: Grow without oxygen and may even be poisoned by it (e.g., *Clostridium botulinum*).

Facultative Anaerobes: Can grow with or without oxygen (e.g., *Escherichia coli*).

4. Reproduction

Bacteria reproduce asexually by binary fission, where a single cell divides into two identical daughter cells. This process is rapid and allows for exponential growth under favorable conditions.

5. Significance

Bacteria play crucial roles in ecosystems:

Decomposers: Break down dead organisms and recycle nutrients.

Nitrogen Fixation: Convert atmospheric nitrogen into forms usable by plants (e.g., *Rhizobium* in legume root nodules).

Bioremediation: Clean up pollutants from the environment.

Some bacteria are pathogenic and cause diseases (e.g., *Streptococcus pneumoniae*). Others are used in biotechnology for producing antibiotics and other products. Bacteria are also utilized in various industries for processes such as fermentation (e.g., yogurt and cheese production) and waste treatment.

Archaea

1. Overview

Archaea are prokaryotic like bacteria but have distinct molecular and biochemical characteristics. Archaea have a single circular DNA molecule and often contain histones associated with their DNA, which is more similar to eukaryotic cells.

2. Structure

Archaea have cell walls that do not contain peptidoglycan. Instead, their cell walls may contain pseudopeptidoglycan, proteins, or polysaccharides. The plasma membranes of archaea contain unique lipids with ether bonds, which differ from the ester bonds found in bacteria and eukaryotes. This structure is thought to provide greater stability in extreme environments. Some archaea possess flagella for movement, but these structures are chemically distinct from bacterial flagella.

3. Classification

Many archaea are extremophiles, living in extreme environments:

Thermophiles: Thrive at high temperatures (e.g., *Thermus aquaticus*).

Halophiles: Prefer high-salt environments (e.g., *Halobacterium salinarum*).

Acidophiles: Live in acidic conditions (e.g., *Ferroplasma acidarmanus*).

Methanogens produce methane as a byproduct of their metabolism and are found in anaerobic environments like swamps and the guts of ruminants. Some archaea are found in more moderate environments, including soil and oceans.

4. Reproduction

Like bacteria, archaea reproduce asexually by binary fission. Their cell division machinery is more similar to eukaryotes than to bacteria.

5. Significance

Archaea play important roles in biogeochemical cycles:

Methane Production: Methanogens contribute to the global carbon cycle by producing methane.

Nutrient Cycling: Involved in nutrient cycles in extreme and non-extreme environments.

Archaea are studied for their unique enzymes and adaptations, which have applications in industrial processes and biotechnology. They are considered to be more closely related to eukaryotes than to bacteria, providing important insights into the evolution of cellular life.

Comparative Analysis of Bacteria and Archaea

Molecular differences between bacteria and archaea include ribosomal RNA and protein compositions. Archaeal ribosomes are more similar to those of eukaryotes. Their cell membranes contain ether-linked lipids, whereas bacterial membranes contain ester-linked lipids. Both domains exhibit a wide range of metabolic diversity, but archaea are notable for their ability to survive in extreme conditions.

Phylogenetic analysis reflects the deep evolutionary separation between Bacteria, Archaea, and Eukarya. Archaea and eukaryotes share a more recent common ancestor compared to Bacteria. Archaea are often found in extreme environments, while bacteria can be found in a wide range of habitats, including extreme conditions and more moderate environments.

Bacteria and Archaea represent two of the three domains of life and are critical to our understanding of microbial diversity and evolution. While both are prokaryotic, they differ significantly in their cellular structures, molecular compositions, and ecological roles. Advances in molecular biology and genomics continue to reveal new insights into their biology, evolutionary relationships, and applications in biotechnology and medicine. Understanding these microorganisms is essential for exploring their roles in ecosystems, their contributions to human health and industry, and their evolutionary significance.

Fungi

Fungi are a diverse group of eukaryotic organisms that play crucial roles in ecosystems as decomposers, symbionts, and pathogens. They are distinct from plants, animals, and bacteria in their cellular structure, metabolism, and reproductive strategies. This chapter explores the characteristics, classification, ecological roles, and significance of fungi.

1. Overview of Fungi

A. Cellular Structure

Cell Wall: Fungi have a rigid cell wall made primarily of chitin, a polysaccharide that provides structural support and protection. This is distinct from the cellulose found in plant cell walls.

Cell Membrane: The plasma membrane of fungi contains ergosterol, a sterol compound similar to cholesterol in animal cells but different from the sterols in plant membranes.

Nucleus and Organelles: Fungal cells have a true nucleus and membrane-bound organelles, including mitochondria, endoplasmic reticulum, and Golgi apparatus.

B. Nutrition

Heterotrophic: Fungi are heterotrophic, meaning they cannot produce their own food and rely on organic compounds from other organisms. They obtain nutrients through extracellular digestion.

Saprophytic: Many fungi decompose dead organic matter, recycling nutrients back into the ecosystem.

Parasitic: Some fungi obtain nutrients from living hosts, often causing disease (e.g., *Candida* species).

Mutualistic: Fungi can form mutualistic relationships with other organisms, such as mycorrhizae with plant roots and lichens with algae or cyanobacteria.

2. Classification of Fungi

Fungi are classified into several major groups based on their reproductive structures and life cycles:

A. Chytridiomycota (Chytrids)

Characteristics: Chytrids are primarily aquatic and have motile spores with flagella, which are unique among fungi.

Reproduction: They reproduce both sexually and asexually. The asexual stage involves the release of motile zoospores.

B. Zygomycota (Zygomycetes)

Characteristics: Zygomycetes are mainly terrestrial and are known for their role in decomposing organic matter.

Reproduction: They reproduce sexually by forming zygospores, which are thick-walled structures resulting from the fusion of gametangia. Asexual reproduction occurs via the production of sporangia that release non-motile spores.

C. Ascomycota (Ascomycetes)

Characteristics: Ascomycetes are the largest group of fungi, including many species with both single-celled and multicellular forms. They are found in various habitats.

Reproduction: Sexual reproduction involves the formation of asci, sac-like structures that contain ascospores. Asexual reproduction often occurs through conidia, which are asexual spores produced on specialized hyphae.

D. Basidiomycota (Basidiomycetes)

Characteristics: Basidiomycetes include mushrooms, puffballs, and shelf fungi. They are characterized by their complex fruiting bodies (basidiocarps) and are important decomposers and mycorrhizal partners.

Reproduction: Sexual reproduction involves the formation of basidia, club-shaped structures where basidiospores are produced. Asexual reproduction is less common but can occur through conidia or budding.

E. Glomeromycota

Characteristics: Glomeromycota are primarily arbuscular mycorrhizal fungi that form symbiotic relationships with plant roots. They are crucial for nutrient uptake in plants.

Reproduction: They reproduce asexually by forming large, multinucleate spores in the soil. Sexual reproduction has not been observed.

3. Reproduction in Fungi

A. Asexual Reproduction

Spore Formation: Asexual reproduction typically involves the production of spores. Spores are dispersed through air, water, or by physical means and can germinate to form new fungal individuals.

Budding: In some fungi, such as yeasts, asexual reproduction occurs through budding, where a new cell grows from the parent cell and eventually separates.

Fragmentation: Some fungi can reproduce by fragmentation, where a part of the fungal body (mycelium) breaks off and grows into a new individual.

B. Sexual Reproduction

Plasmogamy: The fusion of cytoplasm from two compatible fungal cells, leading to a dikaryotic stage where two nuclei coexist in the same cell.

Karyogamy: The fusion of the two nuclei to form a diploid nucleus, which undergoes meiosis to produce haploid spores.

Meiosis: The diploid nucleus undergoes meiosis to produce haploid spores, which are then dispersed and can germinate into new fungal individuals.

4. Ecological Roles of Fungi

A. Decomposers

Fungi play a critical role in decomposing dead organic matter, breaking down complex organic compounds into simpler forms. This process recycles nutrients and contributes to soil fertility.

B. Symbionts

Mycorrhizae: Fungi form mutualistic relationships with plant roots, enhancing nutrient uptake (especially phosphorus) and providing water to plants. In return, fungi receive carbohydrates and other organic compounds from the plant.

Lichens: Lichens are symbiotic associations between fungi and photosynthetic partners (algae or cyanobacteria). They can colonize harsh environments and are important bioindicators of air quality.

C. Pathogens

Some fungi are pathogenic to plants, animals, and humans. Plant pathogens such as *Puccinia* species cause rust diseases, while human pathogens like *Aspergillus* and *Candida* can cause infections, particularly in immunocompromised individuals.

5. Economic and Medical Significance

A. Food Industry

Fermentation: Fungi are used in the production of various fermented foods and beverages, including bread (yeast), beer (brewing yeast), and cheese (molds such as *Penicillium*).

B. Medicine

Antibiotics: Fungi produce antibiotics like penicillin (derived from *Penicillium notatum*) that are used to treat bacterial infections.

Immunosuppressants: Drugs like cyclosporine, derived from *Tolypocladium inflatum*, are used to prevent organ rejection in transplant patients.

C. Biotechnology

Enzyme Production: Fungi produce various enzymes used in industrial processes, such as cellulases for breaking down plant materials and proteases for detergent production.

6. Environmental Impact

Fungi contribute to nutrient cycling, soil formation, and ecosystem dynamics. They are involved in interactions with other organisms and play essential roles in maintaining ecological balance.

Fungi are a diverse and ecologically significant group of eukaryotic organisms with unique structural, metabolic, and reproductive characteristics. Their roles as decomposers, symbionts, and pathogens highlight their importance in various ecosystems and their impact on human health and industry. Understanding fungi enhances our knowledge of biological processes and contributes to advancements in medicine, agriculture, and biotechnology.

Viruses and Prions

Viruses and prions represent two categories of infectious agents that differ significantly from other microorganisms. While they are both capable of causing diseases, their structure, replication mechanisms, and interactions with host organisms are fundamentally different from those of bacteria, fungi, and other life forms. This chapter provides an extensive discussion on the nature, classification, and impact of viruses and prions.

1. Viruses

A. Overview

Viruses are microscopic infectious agents that can infect all types of life forms, from animals and plants to bacteria and archaea. They are unique in that they are not considered living organisms because they cannot carry out metabolic processes or reproduce on their own. Instead, viruses rely on host cells to replicate and propagate.

B. Structure

Viruses exhibit a wide range of structures, but they generally consist of the following components:

Capsid: The protein coat of a virus, known as the capsid, surrounds and protects the viral genome. It is made up of protein subunits called capsomers, which can arrange in various geometric patterns to form shapes such as helical, icosahedral, or complex structures.

Viral Envelope: Some viruses have an outer lipid bilayer known as the envelope, which is derived from the host cell membrane. This envelope may contain viral glycoproteins that facilitate the attachment and entry of the virus into host cells. Non-enveloped viruses only have a capsid.

Genetic Material: The viral genome can be composed of either DNA or RNA. This genetic material may be single-stranded or double-stranded, and it can vary in form (linear, circular, segmented).

C. Classification

Viruses are classified based on several criteria:

Genetic Material: Viruses are categorized by whether their genetic material is DNA or RNA, and whether it is single-stranded or double-stranded. For example, herpesviruses have double-stranded DNA, while influenza viruses have single-stranded RNA.

Capsid Shape: The shape of the capsid contributes to viral classification. For instance, the adenovirus has an icosahedral capsid, while the tobacco mosaic virus has a helical capsid.

Presence of Envelope: Viruses are also classified based on whether they have an envelope. Enveloped viruses include the influenza virus, while non-enveloped viruses include the adenovirus.

Mode of Replication: Different viruses use distinct mechanisms for replication and transcription. For example, retroviruses, such as HIV, use reverse transcription to convert their RNA genome into DNA.

D. Replication Cycle

The replication cycle of a virus involves several steps:

Attachment: The virus binds to specific receptors on the surface of a host cell via viral proteins. This interaction is often highly specific.

Entry: The virus or its genetic material enters the host cell. This can occur through direct fusion with the cell membrane (for enveloped viruses) or endocytosis (for both enveloped and non-enveloped viruses).

Uncoating: The viral capsid is removed, releasing the viral genome into the host cell's cytoplasm or nucleus.

Replication and Transcription: The viral genome is replicated, and viral mRNAs are synthesized. The exact process depends on the type of viral genome (DNA or RNA).

Assembly: New viral particles are assembled from newly synthesized viral genomes and proteins.

Release: New viruses exit the host cell, often by lysis (breaking open the cell) or budding (for enveloped viruses), allowing them to infect new cells.

E. Viral Diseases

Viruses can cause a wide range of diseases, from mild illnesses like the common cold to severe conditions like AIDS and cancer. Examples include:

Influenza: Caused by influenza viruses, leading to respiratory illness.

HIV/AIDS: Caused by the human immunodeficiency virus (HIV), which attacks the immune system.

Hepatitis: Caused by hepatitis viruses, leading to liver inflammation.

Herpes Simplex: Caused by herpes simplex viruses, leading to sores and lesions.

2. Prions

A. Overview

Prions are infectious agents composed solely of misfolded proteins. Unlike viruses, prions lack nucleic acids (DNA or RNA) and are not classified within the traditional biological classification systems. They are responsible for a group of progressive neurodegenerative diseases.

B. Structure and Function

Misfolded Proteins: Prions are abnormal forms of a normal protein, known as the prion protein (PrP). The normal protein is denoted as PrP^C (cellular prion protein), while the misfolded, pathogenic form is PrP^{Sc} (scrapie prion protein). The misfolded proteins can induce normal PrP^C proteins to also misfold, leading to a chain reaction of abnormal protein formation.

No Nucleic Acids: Prions lack DNA or RNA, distinguishing them from viruses and other infectious agents. Their replication involves the conversion of normal proteins into the abnormal prion form.

C. Diseases

Prions cause a range of neurodegenerative diseases known as transmissible spongiform encephalopathies (TSEs). These diseases are characterized by the accumulation of abnormal prion proteins in the brain, leading to neuronal damage and vacuolation. Examples include:

Scrapie: A disease affecting sheep and goats, characterized by itching and neurological symptoms.

Bovine Spongiform Encephalopathy (BSE): Also known as "mad cow disease," it affects cattle and can be transmitted to humans.

Creutzfeldt-Jakob Disease (CJD): A rare, degenerative neurological disorder in humans, with various forms including sporadic, hereditary, and acquired.

Kuru: A disease historically found in Papua New Guinea, transmitted through ritualistic cannibalism.

D. Pathogenesis

Prions cause disease by inducing abnormal folding of normal prion proteins. The accumulation of these abnormal proteins leads to neuronal damage, characterized by brain tissue becoming spongy and losing function. Prion diseases are typically progressive and fatal, with no effective treatment or cure.

E. Transmission

Prions can be transmitted through:

Ingestion: Consuming contaminated meat or products from infected animals.

Medical Procedures: Using contaminated surgical instruments or grafts.

Genetic Transmission: Some prion diseases are inherited due to mutations in the prion protein gene.

Comparative Analysis

Viruses: Require a host cell to replicate, contain genetic material (DNA or RNA), and have a structured capsid or envelope.

Prions: Composed solely of misfolded proteins, lack nucleic acids, and propagate by inducing misfolding of normal proteins.

Viruses and prions represent unique and often devastating classes of infectious agents. Viruses are diverse and complex, with intricate replication mechanisms and a broad range of diseases. Prions, although simpler in structure, cause severe and progressive neurodegenerative diseases through the misfolding of proteins. Understanding these agents is crucial for developing diagnostic, therapeutic, and preventive measures to manage their impact on human and animal health.

Part V: The Plant Kingdom

Chapter 13: Plant Structure and Function

Plant Cells and Tissues

Understanding plant cells and tissues is fundamental to comprehending how plants function and adapt to their environments. Plants have evolved complex structures and specialized tissues that enable them to perform essential functions such as photosynthesis, nutrient transport, and support. This chapter explores the various types of plant cells and tissues, their structures, functions, and roles in the overall physiology of plants.

1. Plant Cells

A. Structure of Plant Cells

Plant cells are eukaryotic and share many similarities with animal cells but also have distinct features:

Cell Wall: Plant cells have a rigid cell wall made of cellulose, hemicellulose, and pectin. This structure provides mechanical support, defines cell shape, and protects against external stresses.

Plasma Membrane: Beneath the cell wall, the plasma membrane controls the movement of substances in and out of the cell. It is composed of a phospholipid bilayer with embedded proteins.

Cytoplasm: The cytoplasm contains various organelles and is the site of many cellular processes.

Nucleus: The nucleus houses the cell's genetic material (DNA) and is involved in regulating gene expression and cell division.

Chloroplasts: These organelles contain chlorophyll and are the site of photosynthesis, where light energy is converted into chemical energy.

Vacuole: Plant cells typically have a large central vacuole that stores nutrients, wastes, and helps maintain turgor pressure, which keeps the cell firm.

Mitochondria: These are the powerhouses of the cell, generating ATP through cellular respiration.

Endoplasmic Reticulum (ER): The ER is involved in the synthesis of proteins (rough ER) and lipids (smooth ER).

Golgi Apparatus: The Golgi apparatus modifies, sorts, and packages proteins and lipids for transport within and outside the cell.

B. Types of Plant Cells

Parenchyma Cells: These are the most common and versatile plant cells, involved in storage, photosynthesis, and tissue repair. They have thin, flexible walls and can divide and differentiate into other cell types.

Collenchyma Cells: These cells provide structural support to growing regions of plants. They have thicker, unevenly thickened cell walls and are often found just beneath the epidermis.

Sclerenchyma Cells: These cells provide strength and support. They have thick, lignified cell walls and are often found in mature regions of plants. They include fibers (long, slender cells) and sclereids (shorter, irregular cells).

Xylem Cells: These cells are involved in the transport of water and minerals from roots to other parts of the plant. They include tracheids (elongated cells with tapered ends) and vessel elements (shorter, wider cells that form continuous tubes).

Phloem Cells: These cells transport the products of photosynthesis (mainly sugars) from source tissues (leaves) to sink tissues (roots and growing areas). They include sieve tube elements (which have perforated sieve plates) and companion cells (which assist in the function of sieve tube elements).

2. Plant Tissues

Plant tissues are categorized into three primary types based on their functions:

A. Dermal Tissue

Epidermis: The outermost layer of cells that provides protection to the plant. It secretes a waxy cuticle that reduces water loss and protects against pathogens. In roots, the epidermis often has root hairs that increase surface area for absorption.

Periderm: Replaces the epidermis in older stems and roots. It consists of cork cells that provide protection and reduce water loss. The periderm also includes the cork cambium, which produces cork cells, and phelloderm, which is a layer of living cells beneath the cork.

B. Vascular Tissue

Xylem: Transports water and dissolved minerals from roots to other parts of the plant. It includes:

Tracheids: Long, tapered cells with thick walls and pits that allow water movement between cells.

Vessel Elements: Shorter, wider cells that form continuous tubes for efficient water transport. Vessel elements have perforated end walls.

Phloem: Transports nutrients, particularly sugars, from photosynthetic tissues to non-photosynthetic tissues. It includes:

Sieve Tube Elements: Cells with sieve plates that allow the flow of phloem sap between cells.

Companion Cells: Cells that are closely associated with sieve tube elements and help in loading and unloading phloem sap.

C. Ground Tissue

Parenchyma: Consists of versatile cells that perform a variety of functions including photosynthesis, storage, and tissue repair. They have thin, flexible cell walls and large vacuoles.

Collenchyma: Provides flexible support to growing tissues. Collenchyma cells have thicker primary cell walls and are typically found in stems and leaves.

Sclerenchyma: Provides rigid support and strength to mature plant parts. It includes:

Fibers: Elongated cells that form bundles, providing strength and flexibility.

Sclereids: Variable in shape, providing hardness to seed coats and fruit skins.

3. Plant Tissue Organization

A. Primary and Secondary Growth

Primary Growth: Occurs at the apical meristems (tips of roots and shoots) and is responsible for the lengthening of stems and roots. It involves the formation of primary tissues and the elongation of plant organs.

Secondary Growth: Occurs in lateral meristems (cambium) and is responsible for the thickening of stems and roots. It involves the production of secondary xylem (wood) and secondary phloem.

B. Tissue Systems

Shoot System: Includes the stems and leaves. The stem supports the plant and contains vascular tissues for transport. Leaves are the primary sites of photosynthesis and gas exchange.

Root System: Anchors the plant and absorbs water and nutrients. Roots also store nutrients and can form symbiotic relationships with fungi (mycorrhizae) for enhanced nutrient uptake.

4. Functional Adaptations

Water Conservation: Many plants have adaptations to minimize water loss, such as thickened cuticles, stomata that open and close in response to environmental conditions, and modified leaves (e.g., spines in cacti).

Nutrient Acquisition: Root adaptations, such as root hairs and mycorrhizal associations, enhance nutrient uptake. Some plants have specialized structures for nutrient acquisition, such as carnivorous plants that trap and digest insects.

Support and Transport: Vascular tissues are adapted to efficiently transport water, minerals, and nutrients throughout the plant. Structural adaptations, such as lignified sclerenchyma cells, provide mechanical support.

Plant cells and tissues are intricately structured and specialized to perform vital functions that support the plant's growth, development, and survival. Understanding the diversity of plant cell types and the organization of plant tissues provides insights into how plants adapt to their environments, manage resources, and contribute to ecosystems. This knowledge is fundamental for fields such as botany, agriculture, and environmental science.

Roots, Stems, and Leaves

Roots, stems, and leaves are fundamental plant organs, each with specialized structures and functions that contribute to the plant's overall growth, development, and survival. This chapter delves into the anatomy and roles of these organs, highlighting their interrelationships and adaptations that enable plants to thrive in diverse environments.

1. Roots

A. Structure of Roots

Roots are essential for anchoring the plant, absorbing water and nutrients, and sometimes storing food. Their structure varies depending on the plant species and environmental conditions:

Root Cap: The root cap is a protective structure at the tip of the root. It covers the growing tip and helps the root navigate through the soil, reducing friction and protecting the meristematic tissue.

Root Meristem: Located just behind the root cap, the root meristem is a region of rapid cell division that contributes to root growth. It produces new cells that differentiate into various root tissues.

Zone of Elongation: Above the root meristem, cells in this zone expand, causing the root to lengthen. This elongation pushes the root deeper into the soil.

Zone of Maturation: In this region, cells differentiate into specialized cell types and begin to perform their specific functions. Root hairs are formed in this zone, increasing the surface area for water and nutrient absorption.

B. Types of Roots

Primary Root: The main root that grows downward from the seedling, often forming a taproot system. It provides stability and accesses deeper water and nutrients. Examples include carrots and dandelions.

Lateral Roots: Roots that branch off from the primary root, increasing the root system's surface area and enhancing nutrient and water uptake.

Fibrous Roots: In fibrous root systems, multiple roots emerge from the base of the stem, creating a dense network of roots. This system is common in monocots like grasses.

Adventitious Roots: Roots that arise from non-root tissues, such as stems or leaves. They are often involved in vegetative reproduction, anchorage, or support. Examples include the aerial roots of orchids and the prop roots of mangroves.

C. Root Functions

Anchorage: Roots anchor the plant securely in the soil, preventing it from being uprooted by wind or water.

Absorption: Root hairs and the root surface are specialized for absorbing water and dissolved nutrients from the soil.

Storage: Roots can store carbohydrates and other nutrients. For instance, beets and sweet potatoes store large amounts of starch in their roots.

Transport: Roots transport water and nutrients to the stem and leaves through vascular tissues.

2. Stems

A. Structure of Stems

Stems support the plant, elevate the leaves for optimal light exposure, and house vascular tissues for nutrient and water transport. Key structural components include:

Nodes and Internodes: Nodes are points on the stem where leaves, branches, or buds emerge. Internodes are the segments between nodes.

Apical Bud: The apical or terminal bud is located at the tip of the stem and is responsible for primary growth in length.

Axillary Buds: Located at the nodes, axillary buds can develop into branches or flowers. They are responsible for secondary growth.

Vascular Bundles: In dicots, vascular bundles (xylem and phloem) are typically arranged in a ring. In monocots, they are scattered throughout the stem. These bundles are crucial for transporting water, nutrients, and sugars.

Cortex: The region between the epidermis and the vascular tissues, consisting of ground tissue that often stores nutrients and provides structural support.

Epidermis: The outer layer of cells that protects the stem and can be covered with a waxy cuticle to reduce water loss.

B. Types of Stems

Herbaceous Stems: Soft, green, and flexible stems found in annuals and perennials. They often die back at the end of the growing season. Examples include sunflowers and lettuce.

Woody Stems: Hard, durable stems with a secondary growth layer of xylem, known as wood. Woody stems are found in trees and shrubs. Examples include oak and maple trees.

Modified Stems: Some stems are modified for specific functions:

Rhizomes: Underground stems that store nutrients and can produce new shoots (e.g., ginger).

Stolons: Horizontal above-ground stems that produce new plants at nodes (e.g., strawberries).

Tubers: Swollen, underground stems that store nutrients (e.g., potatoes).

Cladodes: Flattened stems that perform photosynthesis (e.g., cactus).

C. Stem Functions

Support: Stems provide structural support for leaves, flowers, and fruits, elevating them for better light capture and reproduction.

Transport: Stems contain vascular tissues (xylem and phloem) that transport water, nutrients, and sugars between roots and leaves.

Storage: Some stems store nutrients and water, especially in modified stem structures.

3. Leaves

A. Structure of Leaves

Leaves are the primary sites of photosynthesis, capturing light energy and converting it into chemical energy. Key structures include:

Leaf Blade: The broad, flat part of the leaf that maximizes light absorption. It is often covered by a cuticle to prevent water loss.

Petiole: The stalk that attaches the leaf blade to the stem. It provides flexibility and supports the leaf.

Venation: The pattern of veins in the leaf that includes the central midrib and secondary veins. Venation patterns vary among plant species and are important for structural support and nutrient transport.

Stomata: Small openings on the leaf surface, primarily on the lower side, that allow gas exchange (oxygen, carbon dioxide) and water vapor to move in and out of the leaf. They are surrounded by guard cells that regulate their opening and closing.

Mesophyll: The interior tissue of the leaf, divided into:

Palisade Parenchyma: Cells located under the upper epidermis, rich in chloroplasts, where most photosynthesis occurs.

Spongy Parenchyma: Loosely arranged cells with air spaces that facilitate gas exchange and also participate in photosynthesis.

B. Types of Leaves

Simple Leaves: Have a single, undivided blade. Examples include maple and oak leaves.

Compound Leaves: Have a blade divided into multiple leaflets. Each leaflet has its own petiole, but all are attached to a central rachis. Examples include roses and horse chestnuts.

Modified Leaves: Some leaves are adapted for special functions:

Tendrils: Thin, coiling leaves or leaf parts that help plants climb or support themselves (e.g., peas).

Spines: Modified leaves that reduce water loss and protect against herbivory (e.g., cacti).

Bracts: Modified leaves that are often colorful and attract pollinators (e.g., poinsettias).

C. Leaf Functions

Photosynthesis: Leaves contain chlorophyll and other pigments that capture light energy and convert carbon dioxide and water into glucose and oxygen.

Gas Exchange: Through stomata, leaves facilitate the exchange of gases necessary for photosynthesis and respiration.

Transpiration: The process of water vapor loss from the leaf surface helps with nutrient uptake and cooling of the plant.

Storage: In some plants, leaves store nutrients, water, or energy, such as in succulent plants.

4. Adaptations to Environment

Plants have evolved various adaptations in roots, stems, and leaves to thrive in different environments:

Desert Adaptations: Reduced leaf surface area, thickened cuticles, and specialized water storage tissues help reduce water loss. Examples include succulents and xerophytes.

Aquatic Adaptations: Floating leaves with air spaces and specialized root systems help plants stay afloat and absorb nutrients from water. Examples include water lilies and duckweed.

Arctic Adaptations: Compact growth forms, insulating leaf structures, and antifreeze proteins help plants survive extreme cold. Examples include Arctic willows and snowbells.

Roots, stems, and leaves are integral components of plant structure, each with specialized functions that support the plant's overall health and survival. Understanding the anatomy and adaptations of these organs provides insight into how plants interact with their environment, manage resources, and adapt to various ecological niches. This knowledge is crucial for applications in agriculture, horticulture, and environmental management.

Plant Transport Systems

Plant transport systems are essential for the movement of water, nutrients, and sugars throughout the plant. These systems enable plants to maintain homeostasis, support growth, and respond to environmental changes. This chapter delves into the structures and mechanisms involved in plant transport, focusing on how plants transport water, nutrients, and organic compounds.

1. Vascular Tissues

The plant transport system relies heavily on two main types of vascular tissues: xylem and phloem. These tissues form the plant's vascular system, which functions much like a circulatory system in animals.

A. Xylem

Structure: Xylem tissue is composed of several types of cells:

Tracheids: Elongated cells with tapered ends and thick, lignified secondary walls. They have pits (thin areas of the cell wall) that allow water to move between adjacent cells.

Vessel Elements: Shorter, wider cells that form continuous tubes called vessels. They have perforations (holes) in their end walls, allowing for more efficient water transport compared to tracheids.

Xylem Fibers: Supportive cells with thick walls that provide structural strength.

Xylem Parenchyma: Living cells that store nutrients and contribute to the repair of xylem tissues.

Function: The primary function of xylem is to transport water and dissolved minerals from the roots to the rest of the plant. This movement is driven by transpiration and root pressure.

B. Phloem

Structure: Phloem tissue consists of:

Sieve Tube Elements: Long, tubular cells with perforated end walls (sieve plates) that allow the flow of phloem sap (a mixture of sugars, hormones, and other substances). These cells are alive at maturity but lack a nucleus and many organelles.

Companion Cells: Specialized cells adjacent to sieve tube elements that support their function by loading and unloading sugars and other nutrients into the sieve tubes.

Phloem Fibers: Supportive cells that provide structural strength to the phloem.

Phloem Parenchyma: Cells involved in storage and lateral transport of nutrients.

Function: Phloem transports the products of photosynthesis (primarily sugars) from source tissues (leaves) to sink tissues (roots, fruits, and growing regions). This process, known as translocation, ensures that all parts of the plant receive the nutrients they need for growth and development.

2. Water Transport

A. Mechanisms of Water Transport

Root Pressure: Generated by the active uptake of minerals by root cells, which creates an osmotic gradient that draws water into the roots. This pressure can push water upward into the xylem, though it is not sufficient to explain the full extent of water transport.

Transpiration: The primary mechanism driving water movement in plants. Water evaporates from the stomata in the leaves, creating a negative pressure in the leaf xylem. This negative pressure pulls water from the roots through the xylem vessels in a process called the transpiration-cohesion-tension mechanism.

Cohesion and Adhesion: Water molecules exhibit cohesion (attraction to each other) and adhesion (attraction to the xylem walls). Cohesion helps maintain a continuous water column, while adhesion helps water adhere to the walls of the xylem vessels, reducing the likelihood of air bubbles disrupting the flow.

B. Regulation of Water Loss

Stomata: Small openings on the leaf surface regulated by guard cells. Stomata open to allow gas exchange and close to reduce water loss. The opening and closing are controlled by environmental factors such as light, humidity, and carbon dioxide levels.

Cuticle: A waxy layer on the leaf surface that reduces water loss by minimizing evaporation.

Leaf Modifications: Plants in arid environments may have adaptations such as reduced leaf surface area, thicker cuticles, or modified leaves (e.g., spines in cacti) to conserve water.

3. Nutrient Transport

A. Mineral Nutrient Transport

Uptake: Mineral nutrients are absorbed from the soil through root hairs via active transport or diffusion. Active transport requires energy to move ions against their concentration gradient.

Transport to Xylem: Once inside the root, minerals are transported to the xylem vessels, where they travel with the water stream to the rest of the plant.

Movement in Xylem: Mineral nutrients move with water flow through the xylem to various parts of the plant where they are utilized for growth and development.

B. Organic Compound Transport

Photosynthates: The sugars produced during photosynthesis are transported from the leaves (source) to various parts of the plant (sinks) through the phloem.

Loading and Unloading: Companion cells actively load sugars into the sieve tubes at the source end and unload them at the sink end. This process involves active transport and creates pressure differences that drive the flow of phloem sap.

Phloem Sap Composition: Besides sugars, phloem sap can contain amino acids, hormones, and other organic compounds essential for plant growth and regulation.

4. Integration of Transport Systems

The xylem and phloem are interconnected, working together to maintain plant homeostasis:

Nutrient Cycling: Nutrients absorbed by the roots are transported to the leaves for photosynthesis, and the products are then distributed to other parts of the plant where they are needed for growth, development, and storage.

Water-Nutrient Relationships: The efficiency of water transport affects nutrient uptake and distribution. Proper water availability ensures optimal nutrient transport and plant health.

Stress Responses: Plants can adjust their transport systems in response to environmental stresses. For example, during drought conditions, plants may close their stomata to reduce water loss, affecting the rate of transpiration and nutrient transport.

5. Adaptations in Plant Transport Systems

Plants have evolved various adaptations to optimize their transport systems based on their environment:

Xerophytes: Plants adapted to dry conditions often have deep root systems, thick cuticles, and specialized water storage tissues to minimize water loss and maximize water uptake.

Hydrophytes: Aquatic plants may have large air spaces in their stems and leaves to aid in buoyancy and gas exchange, as well as specialized root structures for nutrient absorption in water.

Halophytes: Plants that grow in saline environments have mechanisms to exclude or excrete excess salts, often through specialized glands or tissues.

The plant transport systems are crucial for maintaining plant health and functionality, allowing for the efficient movement of water, nutrients, and organic compounds. By understanding these systems, we gain insights into how plants adapt to their environments, manage resources, and support their growth and development. This knowledge is essential for applications in agriculture, horticulture, and environmental conservation.

Chapter 14: Plant Physiology

Photosynthesis in Detail

Photosynthesis is a critical biochemical process that converts light energy into chemical energy, producing organic compounds that fuel plant growth and development. It is fundamental not only to plants but to all life on Earth, as it is the primary source of organic matter for almost all organisms and plays a crucial role in the global carbon cycle. This chapter explores the process of photosynthesis in detail, including its stages, key components, and significance.

1. Overview of Photosynthesis

Photosynthesis occurs mainly in the chloroplasts of plant cells and involves the conversion of carbon dioxide (CO_2) and water (H_2O) into glucose ($C_6H_{12}O_6$) and oxygen (O_2) using light energy. The overall equation for photosynthesis can be summarized as:

$$6\ CO_2 + 6\ H_2O + \text{light energy} \rightarrow C_6H_{12}O_6 + 6\ O_2$$

This process can be divided into two main stages: the light-dependent reactions and the light-independent reactions (Calvin cycle).

2. Light-Dependent Reactions

The light-dependent reactions, also known as the photochemical phase, occur in the thylakoid membranes of the chloroplasts. These reactions convert light energy into chemical energy stored in ATP and NADPH, which are then used in the Calvin cycle. Key components and steps include:

A. Photosystems

Photosystem II (PSII): The primary photosystem involved in the light-dependent reactions. It absorbs light energy, which excites electrons and causes the splitting of water molecules (photolysis). This process generates oxygen, protons, and electrons:

$$2\ H_2O \rightarrow O_2 + 4\ H^+ + 4\ e^-$$

Photosystem I (PSI): Absorbs light energy to further excite electrons and reduce NADP+ to NADPH:

$$NADP^+ + 2e^- + 2H^+ \rightarrow NADPH + H_2O$$

B. Electron Transport Chain (ETC)

Plastoquinone (PQ): Transfers electrons from PSII to the cytochrome b6f complex. As electrons move through the chain, protons are pumped into the thylakoid lumen, creating a proton gradient.

Cytochrome b6f Complex: Transfers electrons from plastoquinone to plastocyanin, while pumping protons across the thylakoid membrane.

Plastocyanin (PC): Carries electrons to PSI.

Ferredoxin: Transfers electrons from PSI to NADP+, resulting in the formation of NADPH.

C. ATP Synthesis

Chemiosmosis: The proton gradient created by the electron transport chain drives the synthesis of ATP via ATP synthase. Protons flow through ATP synthase from the thylakoid lumen to the stroma, driving the phosphorylation of ADP to ATP.

3. Light-Independent Reactions (Calvin Cycle)

The Calvin cycle, also known as the dark reactions, occurs in the stroma of the chloroplasts. It utilizes ATP and NADPH produced in the light-dependent reactions to convert CO_2 into glucose. The Calvin cycle comprises three main phases:

A. Carbon Fixation

Ribulose-1,5-bisphosphate (RuBP): Reacts with CO_2 to form a 6-carbon intermediate that immediately splits into two molecules of 3-phosphoglycerate (3-PGA), catalyzed by the enzyme ribulose-1,5-bisphosphate carboxylase/oxygenase (RuBisCO).

B. Reduction

3-Phosphoglycerate (3-PGA): Is phosphorylated by ATP and reduced by NADPH to form glyceraldehyde-3-phosphate (G3P). Some G3P exits the cycle to form glucose and other carbohydrates.

C. Regeneration of RuBP

Glyceraldehyde-3-phosphate (G3P): Using ATP, the remaining G3P molecules are converted back into RuBP, allowing the cycle to continue.

4. Factors Affecting Photosynthesis

Several factors can influence the rate of photosynthesis:

A. Light Intensity

Light Saturation: At low light intensities, photosynthesis rates increase with light intensity. Beyond a certain point, further increases in light do not significantly enhance photosynthesis due to the saturation of photosynthetic pigments and enzymes.

B. Carbon Dioxide Concentration

Carbon Fixation: Higher CO_2 concentrations can increase the rate of photosynthesis, as more CO_2 is available for fixation. However, this effect levels off at high concentrations when other factors become limiting.

C. Temperature

Enzyme Activity: Photosynthesis is temperature-dependent because enzymes involved in the Calvin cycle, such as RuBisCO, have optimal temperature ranges. Extreme temperatures can denature enzymes or slow down metabolic processes.

D. Water Availability

Stomatal Regulation: Water stress can cause stomata to close, reducing CO_2 intake and water loss but also limiting photosynthesis. Adequate water is essential for maintaining stomatal function and overall photosynthetic activity.

5. Photosynthesis and Plant Adaptations

Plants have evolved various adaptations to optimize photosynthesis in different environments:

A. C3 Plants

Typical Photosynthesis: C3 plants, like wheat and rice, fix CO_2 directly in the Calvin cycle. They are efficient under moderate light and CO_2 conditions but can suffer from photorespiration at high temperatures.

B. C4 Plants

Alternative Pathway: C4 plants, such as maize and sugarcane, have an additional carbon fixation step that forms a 4-carbon compound. This adaptation minimizes photorespiration and allows for more efficient photosynthesis under high light and temperature conditions.

C. CAM Plants

Crassulacean Acid Metabolism (CAM): CAM plants, such as succulents and cacti, fix CO_2 at night to reduce water loss. During the day, they use the stored CO_2 for photosynthesis, making them highly efficient in arid environments.

6. Photosynthesis and Global Impact

A. Oxygen Production

Atmospheric Oxygen: Photosynthesis is the primary source of atmospheric oxygen, which is essential for aerobic respiration in most organisms.

B. Carbon Sequestration

Climate Regulation: By absorbing CO_2, photosynthesis helps regulate atmospheric carbon levels and mitigate climate change. Forests, oceans, and other photosynthetic ecosystems play critical roles in carbon sequestration.

C. Food Supply

Agricultural Productivity: Photosynthesis is the foundation of agriculture, providing the energy and organic compounds necessary for crop growth and food production.

Photosynthesis is a complex and vital process that enables plants to convert light energy into chemical energy, producing the organic compounds necessary for their growth and survival. Understanding the details of photosynthesis, including its stages, influencing factors, and adaptations, provides insight into plant physiology and its broader ecological and economic impacts. This knowledge is essential for advancements in agriculture, environmental conservation, and addressing global challenges such as climate change.

Plant Growth and Development

Plant growth and development are intricate processes governed by a combination of genetic factors, environmental conditions, and physiological mechanisms. Understanding these processes is essential for optimizing plant health, improving agricultural yields, and managing plant resources sustainably. This chapter explores the fundamental aspects of plant growth and development, including growth mechanisms, developmental stages, and influencing factors.

1. Plant Growth

Plant growth involves an increase in size and mass due to the accumulation of new cells and tissues. Growth can be categorized into primary and secondary growth.

A. Primary Growth

Location: Occurs at the apical meristems, which are found at the tips of roots and shoots.

Function: Responsible for the elongation of roots and shoots, allowing plants to grow taller and deeper.

Process: Involves the division and differentiation of cells in the meristematic tissues:

Cell Division: New cells are produced by mitosis in the meristematic region.

Cell Elongation: Newly formed cells expand, contributing to the lengthening of the plant.

Cell Differentiation: Elongated cells undergo specialization to form different tissues and organs.

B. Secondary Growth

Location: Occurs in the lateral meristems, primarily the vascular cambium and cork cambium.

Function: Responsible for the increase in girth or diameter of stems and roots, contributing to the overall thickness of woody plants.

Process:

Vascular Cambium: Produces secondary xylem (wood) towards the inside and secondary phloem towards the outside.

Cork Cambium: Produces cork cells that form the outer protective layer of the plant.

2. Plant Development

Plant development encompasses the complex series of changes that lead to the formation of mature plant structures and functions. It includes the transition from embryonic stages to fully developed adult plants.

A. Developmental Stages

Seed Germination

Process: The transition from a dormant seed to an active seedling involves water absorption (imbibition), activation of metabolic processes, and growth.

Phases:

Imbibition: Seed absorbs water, swelling and breaking the seed coat.

Activation: Enzymes are activated, leading to the breakdown of stored nutrients.

Growth: Radicle (embryonic root) and shoot (cotyledons or first leaves) emerge.

Seedling Stage

Characteristics: Rapid growth and establishment of root, stem, and leaf structures.

Development: Formation of primary tissues and initial organ development. The plant establishes its root system and begins photosynthesis.

Vegetative Stage

Characteristics: Continuous growth of leaves, stems, and roots.

Development: The plant develops a more extensive root system and larger foliage to maximize light capture and nutrient absorption.

Reproductive Stage

Characteristics: Formation of reproductive structures such as flowers, fruits, and seeds.

Development: Involves the transition from vegetative growth to reproductive processes, including flowering, pollination, fertilization, and seed development.

Senescence and Dormancy

Characteristics: Aging and eventual death of plant parts or the whole plant.

Development: Involves the breakdown of cellular components and the cessation of metabolic activities. Some plants enter dormancy to survive adverse conditions.

B. Phases of Plant Development

Embryogenesis

Formation of Embryo: The fertilized zygote develops into a mature embryo within the seed, which includes the embryonic root, shoot, and cotyledons.

Seed Coat Formation: Protection of the embryo by the seed coat.

Vegetative Growth

Leaf Development: Formation of leaves for photosynthesis.

Root Development: Establishment of roots for nutrient and water uptake.

Flowering and Reproduction

Flower Formation: Development of floral organs (sepals, petals, stamens, and carpels).

Pollination and Fertilization: Transfer of pollen to the stigma and fusion of gametes.

Fruit and Seed Development

Fruit Formation: Development of ovary into fruit, which houses seeds.

Seed Maturation: Development and ripening of seeds for dispersal.

Senescence and Dormancy

Senescence: Aging process leading to the breakdown of cellular structures.

Dormancy: Period of inactivity to conserve energy and withstand unfavorable conditions.

3. Growth Regulators and Hormones

Plant growth and development are regulated by a variety of hormones and growth regulators, which act as signaling molecules to control various physiological processes.

A. Major Plant Hormones

Auxins

Function: Promote cell elongation, root initiation, and differentiation. Play a role in phototropism (growth towards light) and gravitropism (growth in response to gravity).

Examples: Indole-3-acetic acid (IAA).

Cytokinins

Function: Promote cell division and differentiation, delay senescence, and enhance shoot development.

Examples: Zeatin, kinetin.

Gibberellins

Function: Stimulate stem elongation, seed germination, and flowering. Important for breaking seed dormancy.

Examples: Gibberellic acid (GA).

Abscisic Acid (ABA)

Function: Induces dormancy, regulates water stress responses, and promotes leaf senescence.

Examples: ABA.

Ethylene

Function: Regulates fruit ripening, leaf abscission (drop), and stress responses.

Examples: Ethylene gas.

Brassinosteroids

Function: Promote cell elongation, division, and differentiation. Involved in stress tolerance.

Examples: Brassinolide.

B. Interaction of Hormones

Hormonal Balance: The balance and interaction between different hormones determine the plant's growth responses. For example, auxins and cytokinins work together to control cell division and differentiation in tissues.

4. Environmental Influences on Growth and Development

Plants respond to various environmental factors that influence their growth and development:

A. Light

Phototropism: Growth in response to light direction. Plants grow towards light sources to maximize photosynthesis.

Photoperiodism: Response to the length of day and night. Controls flowering and other developmental processes.

B. Temperature

Thermoperiodism: Response to temperature changes. Optimal temperatures are necessary for enzymatic activities and growth.

Cold Hardiness: Ability to withstand low temperatures. Some plants undergo acclimatization or enter dormancy to survive cold periods.

C. Water Availability

Hydrotropism: Growth towards areas of higher water availability.

Water Stress: Can affect growth, lead to wilting, and trigger stress responses such as stomatal closure and hormone signaling.

D. Soil Nutrients

Nutrient Uptake: Essential for growth and development. Nutrient deficiencies or excesses can impact plant health and productivity.

Soil pH: Affects nutrient availability and root function.

5. Growth and Development in Different Plant Types

A. Herbaceous Plants

Growth Patterns: Typically exhibit rapid growth and short life cycles. Includes annuals and perennials with primary growth.

B. Woody Plants

Growth Patterns: Characterized by secondary growth, leading to an increase in girth. Includes trees and shrubs.

C. Aquatic Plants

Adaptations: Modified structures for buoyancy and gas exchange. Adaptations to low-light conditions in water.

6. Applications in Agriculture and Horticulture

Understanding plant growth and development is crucial for optimizing agricultural practices and improving crop yields:

Crop Breeding: Developing varieties with desirable traits such as improved growth rates, disease resistance, and stress tolerance.

Soil Management: Ensuring proper nutrient and water management to support healthy plant growth.

Plant Propagation: Techniques for producing new plants, including seed germination, cuttings, and tissue culture.

Plant growth and development are dynamic processes influenced by a range of internal and external factors. By understanding the mechanisms and regulators involved, we can better manage plant health, enhance agricultural productivity, and address environmental challenges. This knowledge is essential for advancements in plant science, agriculture, and ecosystem management.

Reproduction in Plants

Plant reproduction is a fundamental process that ensures the continuation of plant species. It can be broadly categorized into asexual (vegetative) reproduction and sexual reproduction. This chapter delves into the mechanisms, structures, and processes involved in both types of reproduction, highlighting their significance, advantages, and roles in plant life cycles.

1. Asexual (Vegetative) Reproduction

Asexual reproduction in plants involves the production of offspring without the involvement of gametes (sex cells). It results in genetically identical clones of the parent plant, which can be advantageous for rapid propagation and maintaining successful traits.

A. Mechanisms of Asexual Reproduction

Vegetative Propagation

Root Suckers: New shoots that arise from the root system of a parent plant. Examples include raspberry and quaking aspen.

Rhizomes: Underground horizontal stems that produce new shoots and roots. Examples include ginger and bamboo.

Tubers: Swollen underground storage organs that produce new plants from buds. Examples include potatoes.

Bulbs: Underground storage organs consisting of a short stem surrounded by fleshy leaves. Examples include onions and tulips.

Corms: Solid, swollen underground stems that store nutrients. Examples include crocuses and gladiolus.

Cloning Techniques

Cuttings: A piece of a plant (stem, leaf, or root) is placed in a suitable medium to develop roots and grow into a new plant. Used extensively in horticulture and agriculture.

Layering: A stem or branch is bent to the ground and covered with soil to encourage root formation. Examples include strawberries and some vines.

Grafting: A segment of one plant (scion) is joined to the rootstock of another plant to propagate desirable traits. Commonly used for fruit trees and ornamental plants.

Special Structures

Runner or Stolons: Horizontal stems that grow along the soil surface and form new plants at nodes. Examples include strawberries and Bermuda grass.

Suckers: Shoots that arise from the base of a plant or from the roots, producing new plants. Examples include many fruit trees and some shrubs.

B. Advantages and Disadvantages of Asexual Reproduction

Advantages:

Rapid Reproduction: Allows for quick spread and colonization of suitable environments.

Genetic Uniformity: Ensures that successful traits are preserved.

No Need for Pollinators: Does not rely on external factors such as wind or insects.

Disadvantages:

Lack of Genetic Diversity: Reduced ability to adapt to changing environmental conditions.

Disease Spread: Susceptibility to diseases and pests that affect the entire clone.

2. Sexual Reproduction

Sexual reproduction involves the fusion of male and female gametes to produce genetically diverse offspring. This process enhances genetic variation and adaptability.

A. Structures Involved in Sexual Reproduction

Flowers

Function: The reproductive organs of angiosperms (flowering plants). Flowers contain both male and female structures and may be unisexual or bisexual.

Parts:

Sepals: Protect the developing flower bud.

Petals: Attract pollinators with their color and scent.

Stamens: Male reproductive organs consisting of the anther (produces pollen) and filament.

Carpels (Pistils): Female reproductive organs consisting of the stigma (receives pollen), style (connects stigma and ovary), and ovary (contains ovules).

Pollination

Process: The transfer of pollen from the anther to the stigma of a flower. Pollination can be mediated by wind, insects, birds, water, or animals.

Types:

Self-Pollination: Pollen from the same flower or plant fertilizes the ovules. Ensures reproduction but limits genetic diversity.

Cross-Pollination: Pollen from one flower fertilizes the ovules of another flower, promoting genetic variation.

Fertilization

Process: The fusion of male and female gametes (sperm and egg cells) results in the formation of a zygote.

Double Fertilization: In angiosperms, one sperm fertilizes the egg to form the zygote, while another sperm fuses with two other cells to form the triploid endosperm, which nourishes the developing embryo.

B. Development of Seeds and Fruits

Seed Development

Embryo Formation: The zygote develops into an embryo, which includes the radicle (future root), shoot, and cotyledons (seed leaves).

Seed Coat Formation: The ovule develops a protective seed coat that encases the embryo and stored nutrients.

Fruit Formation

Ovary Development: The ovary of the flower develops into a fruit, which protects the seeds and aids in their dispersal.

Types of Fruits:

Fleshy Fruits: Have a soft, edible pericarp (e.g., apples, berries).

Dry Fruits: Have a hard or papery pericarp that may split open when mature (e.g., nuts, legumes).

C. Seed Dispersal

Mechanisms: The process by which seeds are spread away from the parent plant to reduce competition and colonize new areas.

Wind Dispersal: Seeds with structures like wings or fluff that facilitate wind transport (e.g., dandelions, maples).

Animal Dispersal: Seeds adhere to fur or are ingested and excreted by animals (e.g., burdock, cherries).

Water Dispersal: Seeds float and are carried by water (e.g., coconuts, water lilies).

Mechanical Dispersal: Seeds are ejected from the fruit when it dries and splits (e.g., peas, violets).

3. Plant Life Cycles

Plants have distinct life cycles that can be categorized into annual, biennial, and perennial types.

A. Annuals

Lifecycle: Complete their life cycle within one growing season. They germinate, grow, flower, produce seeds, and die within a single year.

Examples: Corn, wheat, marigolds.

B. Biennials

Lifecycle: Complete their life cycle over two growing seasons. They typically grow and store energy in the first year and flower and set seeds in the second year.

Examples: Carrots, cabbage.

C. Perennials

Lifecycle: Live for more than two years. They continue to grow and reproduce over multiple years, with some species having dormant periods.

Examples: Trees (oaks, maples), shrubs (roses, blueberries).

4. Reproductive Strategies and Adaptations

Plants have evolved various strategies and adaptations to enhance their reproductive success.

A. Pollination Strategies

Attracting Pollinators: Flowers may produce nectar, emit scents, or have bright colors to attract pollinators.

Timing of Flowering: Flowers may open at specific times of day or seasons to align with pollinator activity.

B. Genetic Diversity

Cross-Pollination: Promotes genetic diversity and adaptability. Plants may have mechanisms to encourage cross-pollination, such as timing of pollen release and stigma receptivity.

C. Seed Dormancy

Adaptation: Seed dormancy allows seeds to survive unfavorable conditions and germinate when conditions are optimal.

Mechanisms: Dormancy can be broken by factors such as temperature changes, water availability, or light.

Reproduction in plants is a diverse and complex process that ensures the survival and adaptation of plant species. Understanding the mechanisms of asexual and sexual reproduction, as well as the various strategies and adaptations, provides insight into plant biology and its applications in agriculture, horticulture, and conservation. By mastering these concepts, we can enhance crop production, manage plant resources effectively, and address environmental challenges.

Chapter 15: Introduction to Animals

Animal Cells and Tissues

Animal cells and tissues form the foundational components of the animal kingdom. Understanding their structure and function is essential for comprehending the complex physiology and organization of animals. This chapter provides an in-depth exploration of animal cell types, tissue organization, and their roles in the overall function of animal bodies.

1. Animal Cells

Animal cells share many features with other eukaryotic cells but also possess unique characteristics suited to their functions in multicellular organisms.

A. Basic Structure

Cell Membrane

Function: Acts as a selective barrier, regulating the movement of substances into and out of the cell.

Structure: Composed of a phospholipid bilayer with embedded proteins, cholesterol, and carbohydrates.

Cytoplasm

Components: Includes the cytosol (fluid portion) and organelles suspended within it.

Function: Site of various metabolic activities and intracellular processes.

Nucleus

Function: Houses genetic material (DNA) and regulates gene expression and cell activities.

Structure: Enclosed by a nuclear envelope with nuclear pores, contains chromatin (DNA and proteins), and a nucleolus (site of ribosome assembly).

Organelles

Mitochondria: Energy production through cellular respiration; known as the powerhouse of the cell.

Endoplasmic Reticulum (ER):

Rough ER: Studded with ribosomes, involved in protein synthesis and processing.

Smooth ER: Lacks ribosomes, involved in lipid synthesis, detoxification, and calcium ion storage.

Golgi Apparatus: Modifies, sorts, and packages proteins and lipids for secretion or delivery to other organelles.

Lysosomes: Contain digestive enzymes for the breakdown of macromolecules and cellular waste.

Peroxisomes: Involved in lipid metabolism and detoxification of harmful substances.

Ribosomes: Sites of protein synthesis, found in the cytoplasm and on the rough ER.

B. Cellular Processes

Cell Division

Mitosis: Process of cell division that results in two genetically identical daughter cells, essential for growth, repair, and asexual reproduction.

Meiosis: Specialized form of cell division that produces four genetically diverse gametes (sperm and eggs) for sexual reproduction.

Cell Communication

Signal Transduction: The process by which cells respond to external signals through receptors and intracellular signaling pathways.

Types: Includes paracrine signaling (local communication), endocrine signaling (hormones), and autocrine signaling (self-regulation).

2. Animal Tissues

Animal tissues are groups of cells with similar structures and functions working together to perform specific tasks. They are categorized into four primary types: epithelial, connective, muscle, and nervous tissues.

A. Epithelial Tissue

Characteristics

Structure: Cells are closely packed with minimal extracellular matrix; form continuous layers.

Function: Covers body surfaces, lines cavities and organs, and forms glands.

Types

Simple Epithelium: Single layer of cells.

Simple Squamous: Thin, flat cells; allows for diffusion and filtration (e.g., alveoli of lungs).

Simple Cuboidal: Cube-shaped cells; involved in secretion and absorption (e.g., kidney tubules).

Simple Columnar: Tall, column-like cells; involved in absorption and secretion (e.g., lining of the digestive tract).

Stratified Epithelium: Multiple layers of cells.

Stratified Squamous: Multiple layers, with surface cells being flat; provides protection (e.g., skin, esophagus).

Stratified Cuboidal and Columnar: Less common; found in glandular ducts and some reproductive tissues.

Specialized Epithelium

Pseudostratified Epithelium: Appears multilayered but is a single layer of varying heights; often ciliated (e.g., respiratory tract).

Transitional Epithelium: Stretchable tissue found in the urinary bladder; accommodates changes in volume.

B. Connective Tissue

Characteristics

Structure: Cells are dispersed in an extracellular matrix; provides structural and metabolic support.

Function: Supports, binds together, and protects tissues and organs.

Types

Loose Connective Tissue: Includes areolar, adipose, and reticular tissues.

Areolar Tissue: Provides elasticity and support; found beneath epithelial tissues.

Adipose Tissue: Stores fat, insulates, and cushions organs (e.g., subcutaneous fat).

Reticular Tissue: Forms supportive networks in lymphoid organs (e.g., spleen, lymph nodes).

Dense Connective Tissue: Includes tendons, ligaments, and dermis.

Dense Regular: Parallel collagen fibers; provides tensile strength (e.g., tendons, ligaments).

Dense Irregular: Randomly arranged collagen fibers; provides strength and elasticity (e.g., dermis of the skin).

Specialized Connective Tissue:

Cartilage: Provides flexible support; includes hyaline (smooth), elastic (flexible), and fibrocartilage (strong) varieties.

Bone: Rigid connective tissue that supports and protects; includes compact and spongy bone.

Blood: Liquid connective tissue involved in transport and immunity; composed of plasma, red blood cells, white blood cells, and platelets.

C. Muscle Tissue

Characteristics

Structure: Cells are elongated and contractile; specialized for movement.

Function: Facilitates movement of the body and its parts.

Types

Skeletal Muscle: Striated, voluntary muscle attached to bones; responsible for body movement (e.g., biceps, quadriceps).

Cardiac Muscle: Striated, involuntary muscle found in the heart; responsible for pumping blood (e.g., heart muscle).

Smooth Muscle: Non-striated, involuntary muscle found in walls of internal organs and blood vessels; controls movements such as digestion and blood flow.

D. Nervous Tissue

Characteristics

Structure: Composed of neurons and supporting glial cells; specialized for communication and processing.

Function: Transmits electrical signals and processes information.

Components

Neurons: Specialized cells that conduct electrical impulses; consist of a cell body (soma), dendrites (receive signals), and an axon (transmits signals).

Glial Cells: Support and protect neurons; include astrocytes, oligodendrocytes, microglia, and Schwann cells.

3. Tissue Organization and Function

Animal tissues are organized into organs and organ systems that perform specific physiological functions.

A. Organ Systems

Integumentary System: Protects the body and helps regulate temperature (e.g., skin, hair, nails).

Skeletal System: Provides support and protection; facilitates movement (e.g., bones, cartilage).

Muscular System: Enables movement and maintains posture (e.g., skeletal muscles, tendons).

Nervous System: Controls and coordinates body activities through electrical signals (e.g., brain, spinal cord, nerves).

Circulatory System: Transports nutrients, gases, and waste products (e.g., heart, blood vessels).

Respiratory System: Facilitates gas exchange (e.g., lungs, trachea).

Digestive System: Breaks down food and absorbs nutrients (e.g., stomach, intestines).

Excretory System: Removes metabolic waste (e.g., kidneys, bladder).

Reproductive System: Facilitates reproduction (e.g., ovaries, testes).

Endocrine System: Regulates bodily functions through hormones (e.g., glands, hormones).

Understanding animal cells and tissues provides a foundation for exploring more complex physiological processes and systems in animals. The structure and function of various cell types and tissues contribute to the overall health, functionality, and adaptability of animals. This knowledge is crucial for fields such as medicine, veterinary science, and comparative biology.

Animal Body Plans and Symmetry

Understanding animal body plans and symmetry is essential for classifying animals and understanding their evolutionary relationships, functional adaptations, and ecological roles. This chapter explores the different body plans and types of symmetry found in the animal kingdom, highlighting their significance and variations across different phyla.

1. Body Plans

Animal body plans refer to the general architectural layout of an animal's body, including its organization of tissues and organs. The body plan determines how an organism interacts with its environment and how it performs various physiological functions.

A. Symmetry

Symmetry is a key aspect of body plan organization, and it reflects how the body parts are arranged relative to an axis or central point. There are three primary types of symmetry:

Asymmetry

Definition: Lack of a symmetrical body plan; no discernible symmetry.

Examples: Sponges (Phylum Porifera). Sponges have a porous body structure and lack a defined shape, with their cells organized in a loose, non-symmetrical arrangement.

Radial Symmetry

Definition: Body parts are arranged around a central axis, so that any longitudinal cut through the center produces mirror-image halves.

Characteristics: Typically found in organisms that are sessile (fixed in one place) or slow-moving, as this symmetry allows them to interact with their environment from all directions.

Examples:

Cnidarians: Including jellyfish, sea anemones, and corals. These animals have a central oral-aboral axis, with their body parts arranged around it.

Echinoderms: Such as starfish and sea urchins, which exhibit pentamerous radial symmetry (five-fold symmetry).

Bilateral Symmetry

Definition: The body has a single plane of symmetry, resulting in two mirror-image halves along a central axis.

Characteristics: This type of symmetry is associated with more complex, active animals that move in a directed manner. It allows for the development of a distinct head (cephalization) and a more streamlined body.

Examples:

Arthropods: Including insects, spiders, and crustaceans. They have a segmented body with bilateral symmetry.

Chordates: Such as vertebrates (mammals, birds, reptiles) and their relatives. They exhibit bilateral symmetry with a distinct head and tail.

B. Body Cavity Types

The body cavity, or coelom, is the fluid-filled space within the body that separates the digestive tract from the outer body wall. It provides space for the development and movement of organs and is important for the classification of animals.

Acoelomates

Definition: Animals without a body cavity; their tissues are packed closely together.

Characteristics: Typically have a solid body and a digestive cavity that is not separated from the outer body wall.

Examples: Flatworms (Phylum Platyhelminthes).

Pseudocoelomates

Definition: Animals with a pseudocoel, a body cavity that is not fully lined with mesodermal tissue.

Characteristics: The pseudocoel acts as a hydrostatic skeleton and facilitates the movement of internal organs.

Examples: Roundworms (Phylum Nematoda).

Coelomates (Eucoelomates)

Definition: Animals with a true coelom, a body cavity fully lined with mesodermal tissue.

Characteristics: Allows for more complex organ development and greater mobility.

Examples: Annelids (segmented worms), arthropods, mollusks, and chordates.

2. Developmental Patterns

The developmental pattern of an animal can influence its body plan and symmetry. Two major developmental patterns are:

A. Protostome Development

Characteristics:

Cleavage: Spiral and determinate, meaning the fate of each cell is determined early in development.

Coelom Formation: Occurs through schizocoely, where the coelom forms by splitting the mesodermal mass.

Blastopore Fate: The blastopore (initial opening of the embryo) becomes the mouth.

Examples:

Phyla: Arthropoda (insects, spiders), Mollusca (snails, clams), Annelida (earthworms, leeches).

B. Deuterostome Development

Characteristics:

Cleavage: Radial and indeterminate, meaning the fate of the cells is not fixed until later in development.

Coelom Formation: Occurs through enterocoely, where the coelom forms from outpocketings of the embryonic gut.

Blastopore Fate: The blastopore becomes the anus, and the mouth forms later.

Examples:

Phyla: Chordata (vertebrates), Echinodermata (starfish, sea urchins).

3. Evolutionary Significance

The variations in body plans and symmetry reflect the evolutionary adaptations of animals to their environments and lifestyles. These adaptations influence:

A. Locomotion

Radial Symmetry: Suited for sedentary or slow-moving organisms that need to interact with the environment from all directions.

Bilateral Symmetry: Facilitates directional movement and complex behaviors, allowing for better exploration and resource acquisition.

B. Organ Development

Acoelomates: Simple body structures with minimal internal complexity.

Pseudocoelomates: Intermediate complexity, with a flexible body cavity aiding in organ development.

Coelomates: Advanced organ systems and greater body complexity, allowing for specialized functions and increased mobility.

4. Comparative Anatomy and Phylogeny

Comparing body plans and symmetry among different animal phyla provides insights into their evolutionary relationships and adaptations:

Phylum Porifera (Sponges): Asymmetrical body plan with a porous structure; lacks true tissues and organs.

Phylum Cnidaria (Jellyfish, Corals): Radial symmetry with specialized stinging cells (cnidocytes); simple digestive cavity.

Phylum Platyhelminthes (Flatworms): Bilateral symmetry with aacoelomate body plan; simple organ systems.

Phylum Arthropoda (Insects, Spiders): Bilateral symmetry with a coelomate body plan; segmented body and exoskeleton.

Phylum Chordata (Vertebrates): Bilateral symmetry with a coelomate body plan; advanced organ systems and cephalization.

Animal body plans and symmetry are fundamental concepts in understanding animal biology. They provide insight into the structural and functional adaptations of different animal groups, reflecting their evolutionary history and ecological roles. Studying these aspects helps us appreciate the diversity of life forms and their specialized adaptations, contributing to our knowledge of biology and the natural world.

Worms and Mollusks

Invertebrates are a diverse group of animals that lack a backbone, and they include a wide variety of species with distinct body plans and ecological roles. This chapter focuses on two significant invertebrate phyla: **worms** and **mollusks**. Each group exhibits unique characteristics and adaptations that allow them to thrive in various environments.

1. Worms

Worms belong to several phyla within the invertebrates, characterized by their elongated, soft bodies. For simplicity, we'll discuss the three major phyla of worms: **Platyhelminthes (flatworms)**, **Nematoda (roundworms)**, and **Annelida (segmented worms)**.

A. Phylum Platyhelminthes (Flatworms)

General Characteristics

Body Plan: Bilateral symmetry, acoelomate (no body cavity between the digestive tract and outer body wall).

Body Structure: Flattened dorsoventrally, which increases surface area for gas exchange and nutrient absorption.

Digestive System: Incomplete; a single opening serves as both mouth and anus.

Nervous System: Simple, with a pair of cerebral ganglia (brain-like structures) and nerve cords.

Types

Free-Living Flatworms: Such as planarians (e.g., *Dugesia*). They are mostly aquatic or terrestrial and exhibit a simple digestive system and sensory organs.

Parasitic Flatworms: Includes **trematodes (flukes)** and **cestodes (tapeworms)**.

Trematodes: Often have complex life cycles involving multiple hosts (e.g., liver flukes, blood flukes). They attach to hosts using suckers and are known for causing diseases such as schistosomiasis.

Cestodes: Long, segmented parasites with a head structure (scolex) equipped with hooks or suckers for attachment to the host's intestine. They include species like *Taenia* (tapeworms), which can cause digestive issues in humans.

Reproduction

Asexual Reproduction: Many flatworms can regenerate lost body parts and reproduce asexually through fragmentation.

Sexual Reproduction: Most are hermaphroditic, possessing both male and female reproductive organs, which allows for cross-fertilization.

B. Phylum Nematoda (Roundworms)

General Characteristics

Body Plan: Bilateral symmetry, pseudocoelomate (body cavity is not fully lined by mesoderm).

Body Structure: Cylindrical, with a smooth, unsegmented body covered by a tough cuticle that is periodically molted.

Digestive System: Complete, with separate mouth and anus, allowing for continuous digestion.

Nervous System: Simple, with a nerve ring around the pharynx and longitudinal nerve cords.

Types

Free-Living Nematodes: Found in soil and aquatic environments, they play important roles in decomposition and nutrient cycling (e.g., *Caenorhabditis elegans*).

Parasitic Nematodes: Includes species that cause diseases in humans and animals.

Ascarids: Such as *Ascaris lumbricoides*, which can cause intestinal infections.

Hookworms: E.g., *Ancylostoma* and *Necator*, which attach to the host's intestinal wall and feed on blood.

Filarial Worms: Such as *Wuchereria bancrofti*, which cause lymphatic filariasis or elephantiasis.

Reproduction

Sexual Reproduction: Most nematodes have separate sexes and reproduce through internal fertilization. Female nematodes can produce thousands of eggs.

C. Phylum Annelida (Segmented Worms)

General Characteristics

Body Plan: Bilateral symmetry, coelomate (true body cavity lined with mesoderm).

Body Structure: Segmented, with each segment containing similar sets of organs, contributing to more complex structures and functions.

Digestive System: Complete, with a specialized mouth, pharynx, esophagus, crop (for food storage), gizzard (for grinding), and intestine.

Nervous System: More developed, with a ventral nerve cord and segmental ganglia.

Types

Oligochaetes: Such as earthworms (*Lumbricus terrestris*). They are primarily terrestrial and play crucial roles in soil aeration and nutrient recycling.

Polychaetes: Mostly marine, with many segments and parapodia (bristle-like structures) for locomotion and respiration (e.g., *Nereis*). They are often colorful and have diverse ecological roles.

Hirudinea (Leeches): Includes species like the medicinal leech (*Hirudo medicinalis*). They are mostly freshwater and can be parasitic or predatory, using suckers for attachment and feeding.

Reproduction

Asexual Reproduction: Some annelids, like earthworms, can reproduce asexually through fragmentation.

Sexual Reproduction: Most are hermaphroditic, with both reproductive organs in one individual. They engage in cross-fertilization.

2. Mollusks

Mollusks are a diverse phylum of soft-bodied animals, many of which have a hard shell. They exhibit a wide range of body plans and adaptations.

A. General Characteristics

Body Plan

Mantle: A significant part of the body wall that secretes the shell (if present) and forms the mantle cavity.

Foot: A muscular organ used for locomotion, attachment, or feeding.

Visceral Mass: Contains the internal organs, including the digestive, reproductive, and excretory systems.

Digestive System: Complete, with a specialized radula (in most) for feeding, which is a toothed, chitinous structure used to scrape or cut food.

Circulatory System: Most mollusks have an open circulatory system (except cephalopods), where blood is not always confined to vessels and bathes the organs directly.

Respiratory System: Can include gills (in aquatic species) or lungs (in terrestrial species).

B. Major Classes of Mollusks

Class Gastropoda (Gastropods)

Characteristics: Includes snails and slugs. They have a single, often spirally coiled shell (if present) and a distinct head with sensory organs.

Examples:

Land Snails: Such as *Cornu aspersum*, which have adapted to terrestrial environments.

Marine Snails: Such as *Conus* species, known for their venomous stings.

Slugs: Lack a prominent shell and are found in moist environments.

Class Bivalvia (Bivalves)

Characteristics: Includes clams, oysters, and mussels. They have a two-part shell and a body that is laterally compressed.

Examples:

Clams: Such as *Mercenaria mercenaria*, which burrow into sand or mud.

Oysters: Like *Crassostrea* species, which are important for reef building and filtration.

Class Cephalopoda (Cephalopods)

Characteristics: Includes squids, octopuses, and cuttlefish. They have highly developed nervous systems, complex eyes, and are known for their intelligence and camouflage abilities.

Examples:

Octopuses: Such as *Octopus vulgaris*, known for their problem-solving abilities and ability to change color and texture.

Squids: Including the giant squid (*Architeuthis*), which can grow to impressive sizes and have a streamlined body for fast swimming.

Class Polyplacophora (Chitons)

Characteristics: Marine mollusks with eight overlapping plates along their dorsal side, providing protection and flexibility.

Examples: Chitons like *Katharina* species, which cling to rocks in intertidal zones.

C. Reproduction and Development

Sexual Reproduction: Most mollusks reproduce sexually with separate sexes, but some, such as many gastropods, are hermaphroditic.

External Fertilization: Common in many marine species (e.g., bivalves, some gastropods).

Internal Fertilization: Found in most terrestrial and some marine species (e.g., cephalopods).

Developmental Stages: Mollusks often undergo a trochophore larval stage and then a veliger stage before reaching adulthood. Development varies among classes:

Direct Development: In cephalopods, where juveniles resemble adults.

Indirect Development: In bivalves and some gastropods, with larval stages that differ significantly from adults.

3. Ecological and Economic Importance

A. Ecological Roles

Nutrient Cycling: Worms, especially earthworms, play a crucial role in soil aeration and nutrient recycling.

Food Source: Both worms and mollusks are essential food sources for many animals, including birds, fish, and humans.

Habitat Formation: Mollusks such as corals and oysters build reefs that provide habitats for diverse marine life.

B. Economic Importance

Aquaculture: Mollusks such as oysters, clams, and mussels are farmed for human consumption and contribute significantly to global seafood markets.

Medicinal and Industrial Uses: Certain mollusks and worms are used in medicine, research, and industry (e.g., leeches for medical purposes, mollusk shells for calcium supplements).

Worms and mollusks represent two diverse and ecologically significant groups within the invertebrates. Their varied body plans, adaptations, and roles in ecosystems underscore their evolutionary success and importance. Understanding their biology not only enhances our knowledge of invertebrate diversity but also highlights their critical contributions to both natural environments and human economies.

Arthropods and Echinoderms

Invertebrates are a diverse group of animals that lack a backbone, and among them, arthropods and echinoderms are two highly successful and ecologically important phyla. This chapter delves into their characteristics, classification, and roles in the environment.

1. Arthropods

Arthropods are the largest and most diverse group of animals on Earth, comprising over a million described species. They are characterized by their segmented bodies, exoskeletons, and jointed appendages.

A. General Characteristics

Exoskeleton

Composition: Made of chitin, a tough, flexible polysaccharide that provides protection and support.

Function: Protects against physical damage and dehydration but must be periodically molted (ecdysis) to allow for growth.

Segmentation

Body Segments: Arthropods have segmented bodies that are often divided into head, thorax, and abdomen (in insects) or cephalothorax and abdomen (in spiders and crustaceans).

Functional Specialization: Segments may be specialized for different functions, such as locomotion, feeding, or reproduction.

Jointed Appendages

Types: Include legs, antennae, and mouthparts.

Function: Adapted for various functions like walking, sensing the environment, or manipulating food.

Respiratory System

Types:

Tracheal System: Insects have a network of tubes that deliver oxygen directly to tissues.

Gills: Aquatic arthropods, such as crustaceans, use gills for gas exchange.

Book Lungs: In some arachnids, such as spiders, book lungs are used for gas exchange.

Circulatory System

Type: Open circulatory system where hemolymph (a fluid equivalent to blood) is pumped into the hemocoel and bathes the organs directly.

B. Major Classes of Arthropods

Class Insecta (Insects)

General Characteristics: Three pairs of legs, one pair of antennae, and typically two pairs of wings.

Diversity: Includes beetles, butterflies, ants, and flies.

Ecological Roles: Pollinators, decomposers, and prey for other animals. Some insects are also pests or vectors for diseases.

Class Arachnida (Arachnids)

General Characteristics: Four pairs of legs, no antennae, and typically have a cephalothorax and abdomen.

Examples: Spiders, scorpions, ticks, and mites.

Ecological Roles: Predators (e.g., spiders controlling insect populations), parasites (e.g., ticks transmitting diseases).

Class Crustacea (Crustaceans)

General Characteristics: Primarily aquatic with varying numbers of legs and often have two pairs of antennae.

Examples: Crabs, lobsters, shrimp, and barnacles.

Ecological Roles: Important in marine and freshwater ecosystems as detritivores, grazers, and prey for larger animals.

Class Myriapoda (Myriapods)

General Characteristics: Numerous body segments, each with one or two pairs of legs.

Types: Includes centipedes (one pair of legs per segment) and millipedes (two pairs of legs per segment).

Ecological Roles: Detritivores, contributing to soil health by breaking down organic matter.

C. Development and Reproduction

Metamorphosis

Complete Metamorphosis: Insects undergo four life stages: egg, larva, pupa, and adult (e.g., butterflies, beetles).

Incomplete Metamorphosis: Insects develop through three stages: egg, nymph, and adult (e.g., grasshoppers, cockroaches).

Reproduction

Sexual Reproduction: Most arthropods reproduce sexually with separate sexes, although some can reproduce asexually.

Fertilization: Internal fertilization is common, with males transferring sperm to females through specialized structures.

2. Echinoderms

Echinoderms are marine invertebrates known for their unique radial symmetry, calcite endoskeleton, and water vascular system. They are exclusively found in marine environments.

A. General Characteristics

Radial Symmetry

Type: Pentamerous radial symmetry, usually with five or more radiating parts (e.g., arms in starfish).

Function: Allows echinoderms to interact with their environment from multiple directions, which is advantageous for their sessile or slow-moving lifestyles.

Endoskeleton

Composition: Made of calcareous plates or ossicles embedded in the skin, providing support and protection.

Structure: Can be spiny or smooth depending on the species.

Water Vascular System

Function: A network of hydraulic canals used for locomotion, feeding, and respiration.

Components: Includes tube feet, which are small, flexible appendages operated by hydraulic pressure.

Digestive System

Type: Complete, with a mouth located on the underside and an anus (when present) on the top side.

Feeding: Methods vary among species, including filter feeding, grazing, and predation.

B. Major Classes of Echinoderms

Class Asteroidea (Starfish or Sea Stars)

Characteristics: Five or more arms extending from a central disc; equipped with tube feet for movement and feeding.

Examples: *Asterias* species, known for their ability to regenerate lost arms.

Class Ophiuroidea (Brittle Stars and Basket Stars)

Characteristics: Central disc with long, flexible arms that are distinct and separated from the central body.

Movement: Use their arms for locomotion and capturing prey.

Class Echinoidea (Sea Urchins and Sand Dollars)

Characteristics: Spherical or flattened bodies covered with spines; have a hard, fused exoskeleton called a test.

Examples: *Strongylocentrotus* (sea urchins) and *Dendraster* (sand dollars).

Class Holothuroidea (Sea Cucumbers)

Characteristics: Elongated, soft-bodied with a leathery skin; tube feet are typically located on the ventral side.

Feeding: Most are filter feeders or detritivores.

Class Crinoidea (Sea Lilies and Feather Stars)

Characteristics: Cup-shaped body with long, feathery arms used for filter feeding.

Habitat: Mostly sessile and attached to substrates in the deep sea.

C. Reproduction and Development

Asexual Reproduction

Regeneration: Many echinoderms can reproduce asexually through regeneration, where lost parts (e.g., arms) can grow into new individuals.

Sexual Reproduction

External Fertilization: Most echinoderms reproduce by releasing eggs and sperm into the water, where fertilization occurs externally.

Larval Stages: Typically include a free-swimming larval stage before settling into the adult form.

3. Ecological and Economic Importance

A. Ecological Roles

Arthropods: Serve various roles, including pollination, decomposition, and as a food source for many other organisms. They also impact agriculture, ecosystems, and human health through their interactions with plants and animals.

Echinoderms: Contribute to marine ecosystems by participating in nutrient cycling, forming coral reef structures, and controlling algal populations. They also play roles in benthic communities.

B. Economic Importance

Arthropods: Include agricultural pests (e.g., locusts, aphids) and beneficial insects (e.g., bees, ladybugs). They also have roles in medicine (e.g., insect venom for research) and industry (e.g., silk production).

Echinoderms: Some species, like sea cucumbers, are harvested for food and traditional medicine in various cultures. Echinoderms also contribute to marine biodiversity, which supports tourism and recreation.

Arthropods and echinoderms represent two of the most diverse and ecologically significant phyla of invertebrates. Arthropods, with their segmented bodies and jointed appendages, have adapted to nearly every habitat on Earth. Echinoderms, with their radial symmetry and unique water vascular system, play crucial roles in marine ecosystems. Understanding their biology, ecology, and economic importance provides valuable insights into their roles in the natural world and their interactions with human activities.

Chapter 17: Vertebrates
Fish, Amphibians, and Reptiles

Vertebrates are animals with a backbone or spinal column, and they are part of the subphylum Vertebrata within the phylum Chordata. This chapter explores the diverse classes of vertebrates, focusing on fish, amphibians, and reptiles. Each group represents a significant evolutionary step and showcases unique adaptations that allow them to thrive in various environments.

1. Fish

Fish are aquatic vertebrates characterized by their gills, fins, and scales. They represent the earliest vertebrates and are divided into three main classes: **Agnatha (jawless fish)**, **Chondrichthyes (cartilaginous fish)**, and **Osteichthyes (bony fish)**.

A. Class Agnatha (Jawless Fish)

General Characteristics

Body Structure: Eel-like bodies with smooth, scaleless skin.

Skeleton: Cartilaginous skeleton, lacking jaws but possessing a sucker-like mouth.

Respiration: Gills with multiple openings.

Examples

Lampreys: Eel-shaped, parasitic or scavenging fish with a toothed, funnel-shaped mouth.

Hagfish: Slime-producing fish with a cartilaginous skull and no true vertebrae.

Ecological Roles: Both lampreys and hagfish are important in their ecosystems as scavengers and parasites. Lampreys can impact fish populations in freshwater systems, while hagfish play a role in deep-sea ecosystems.

B. Class Chondrichthyes (Cartilaginous Fish)

General Characteristics

Body Structure: Predominantly cartilaginous skeleton with movable jaws and paired fins.

Scales: Placoid scales, which are small, tooth-like structures.

Respiration: Typically have 5-7 gill slits.

Examples

Sharks: Predatory fish with streamlined bodies and well-developed senses, including the ability to detect electrical fields.

Rays and Skates: Flattened bodies adapted for life on the sea floor, with pectoral fins modified into wings.

Ecological Roles: Cartilaginous fish are top predators in many marine ecosystems, controlling populations of other marine species and maintaining ecological balance. Rays and skates contribute to benthic ecosystems by feeding on invertebrates.

C. Class Osteichthyes (Bony Fish)

General Characteristics

Body Structure: Bony skeleton, usually covered with cycloid or ctenoid scales.

Swim Bladder: Gas-filled organ that provides buoyancy.

Respiration: Gills with a bony operculum (gill cover) for protection and efficient water flow.

Examples

Ray-Finned Fish: Such as salmon, trout, and tuna. Characterized by fins supported by thin, flexible rays.

Lobed-Finned Fish: Such as coelacanths and lungfish, with fleshy, lobed fins that are more robust and resemble the limbs of terrestrial vertebrates.

Ecological Roles: Bony fish are incredibly diverse and occupy nearly every aquatic habitat. They serve as important prey for larger animals and are crucial for maintaining the health of aquatic ecosystems.

2. Amphibians

Amphibians are vertebrates that typically live both in water and on land during different stages of their life cycle. They include three main orders: **Anura (frogs and toads)**, **Caudata (salamanders and newts)**, and **Gymnophiona (caecilians)**.

A. Order Anura (Frogs and Toads)

General Characteristics

Body Structure: Compact bodies with long hind limbs adapted for jumping, and a wide mouth.

Skin: Moist, glandular skin that is involved in respiration and water balance.

Life Cycle: Undergo metamorphosis from aquatic larvae (tadpoles) to terrestrial adults.

Examples

Frogs: Such as *Rana* species, which are known for their jumping ability and vocalizations.

Toads: Like *Bufo* species, which have warty skin and are more terrestrial.

Ecological Roles: Frogs and toads are important as both predators of insects and prey for many other animals. They also serve as indicators of environmental health due to their sensitivity to changes in their habitat.

B. Order Caudata (Salamanders and Newts)

General Characteristics

Body Structure: Elongated bodies with tails and limbs of roughly equal size.

Skin: Smooth, often moist, and used for cutaneous respiration.

Life Cycle: Typically have an aquatic larval stage and a terrestrial adult stage, though some species are fully aquatic.

Examples

Salamanders: Such as *Ambystoma* species, which can be fully terrestrial or aquatic.

Newts: Like *Notophthalmus* species, which have a semi-aquatic lifestyle and can undergo significant color changes during their life cycle.

Ecological Roles: Salamanders and newts help control insect populations and serve as prey for a variety of predators. They also play a role in maintaining the health of their aquatic and terrestrial ecosystems.

C. Order Gymnophiona (Caecilians)

General Characteristics

Body Structure: Worm-like, limbless amphibians with a cylindrical body covered in smooth skin.

Skin: Moist and often used for respiration.

Habitat: Primarily burrowers, living in moist soil or leaf litter.

Examples

Caecilians: Such as *Ichthyophis* species, which are mostly blind and adapted to a subterranean lifestyle.

Ecological Roles: Caecilians are important in their ecosystems as insectivores and contribute to soil health through their burrowing activities.

3. Reptiles

Reptiles are cold-blooded vertebrates with a dry, scaly skin and a fully adapted terrestrial life cycle. They are divided into four main groups: **Chelonia (turtles and tortoises)**, **Squamata (lizards and snakes)**, **Crocodylia (crocodiles and alligators)**, and **Rhynchocephalia (tuataras)**.

A. Order Chelonia (Turtles and Tortoises)

General Characteristics

Body Structure: Enclosed in a bony shell made up of a dorsal carapace and a ventral plastron.

Limbs: Adapted for various habitats; aquatic turtles have webbed feet, while tortoises have sturdy, club-like legs.

Examples

Turtles: Such as *Chelonia mydas* (green sea turtle), which are mostly aquatic.

Tortoises: Like *Testudo* species, which are primarily terrestrial.

Ecological Roles: Turtles and tortoises contribute to their ecosystems by dispersing seeds, controlling aquatic vegetation, and serving as prey for larger animals.

B. Order Squamata (Lizards and Snakes)

General Characteristics

Body Structure: Lizards have four legs and a well-developed tail, while snakes are limbless with elongated bodies.

Scales: Overlapping, keratinized scales that provide protection and reduce water loss.

Examples

Lizards: Such as *Iguana* species, which are often arboreal or terrestrial.

Snakes: Including *Naja* (cobras) and *Agkistrodon* (pit vipers), which are diverse in their diet and habitat.

Ecological Roles: Lizards and snakes are important as predators of insects, rodents, and other small animals, and they also serve as prey for birds and mammals. They play crucial roles in maintaining ecological balance.

C. Order Crocodylia (Crocodiles and Alligators)

General Characteristics

Body Structure: Large, semi-aquatic reptiles with long, powerful tails and a robust, armored body.

Teeth: Conical teeth adapted for grabbing and crushing prey.

Examples

Crocodiles: Such as *Crocodylus porosus* (saltwater crocodile), which are found in tropical regions.

Alligators: Like *Alligator mississippiensis* (American alligator), which are primarily found in freshwater habitats.

Ecological Roles: Crocodiles and alligators are apex predators, controlling populations of other animals and influencing the structure of aquatic ecosystems. Their nesting behaviors also create habitats for other species.

D. Order Rhynchocephalia (Tuataras)

General Characteristics

Body Structure: Ancient reptiles with a spiny crest along the back and a unique dentition.

Habitat: Native to New Zealand, tuataras are adapted to cooler climates compared to other reptiles.

Examples

Tuataras: *Sphenodon punctatus*, the only surviving species, is considered a living fossil due to its ancient lineage.

Ecological Roles: Tuataras help control insect populations and serve as prey for larger predators. Their unique evolutionary history provides valuable insights into reptilian evolution.

4. Ecological and Economic Importance

A. Ecological Roles

Fish: Serve as the foundation of aquatic food webs, influencing the structure of marine and freshwater ecosystems.

Amphibians: Act as indicators of environmental health and play roles in both terrestrial and aquatic ecosystems.

Reptiles: Contribute to maintaining ecological balance as predators and prey, and their behaviors can impact their habitats.

B. Economic Importance

Fish: Important for commercial and recreational fishing, as well as for aquaculture industries.

Amphibians: Their skin secretions are used in medical research, and they also have roles in pest control.

Reptiles: Include species that are valuable for the pet trade, as well as those used in traditional medicine and research.

Fish, amphibians, and reptiles each represent distinct evolutionary paths and showcase a range of adaptations that allow them to thrive in their environments. From the aquatic diversity of fish to the dual-life of amphibians and the terrestrial adaptations of reptiles, these vertebrate groups illustrate the complexity and adaptability of life on Earth. Understanding their biology, ecology, and economic significance helps appreciate their roles in natural ecosystems and human societies.

Birds and Mammals

Birds and mammals are two highly specialized and diverse groups of vertebrates, characterized by unique adaptations that allow them to inhabit a wide range of environments. This chapter explores their key features, evolutionary advancements, and ecological and economic significance.

1. Birds

Birds (Class Aves) are warm-blooded vertebrates known for their feathers, beaks, and hollow bones, which contribute to their ability to fly. They are incredibly diverse, with over 10,000 species adapted to various ecological niches.

A. General Characteristics

Feathers

Structure: Made of keratin, feathers provide insulation, aid in flight, and play a role in courtship displays.

Types: Include contour feathers (for aerodynamics and insulation), down feathers (for insulation), and flight feathers (for propulsion).

Skeleton

Composition: Lightweight and reinforced with air sacs, reducing body weight for flight.

Structure: Many bones are hollow, and the sternum (breastbone) is often enlarged to anchor powerful flight muscles.

Respiratory System

Air Sacs: Birds have a unique respiratory system with air sacs that allow for a continuous flow of air through the lungs, providing efficient gas exchange during both inhalation and exhalation.

Beak

Function: Adapted to various feeding habits, from insectivory to piscivory, and replaces teeth, which are absent in birds.

Reproduction

Eggs: Birds lay hard-shelled eggs, which provide protection and support for the developing embryo. Parental care is common, with many species exhibiting complex nesting behaviors.

B. Major Orders of Birds

Order Passeriformes (Perching Birds)

Characteristics: Largest and most diverse bird order, characterized by a flexible toe arrangement that facilitates perching.

Examples: Sparrows, robins, and finches.

Order Accipitriformes (Raptors)

Characteristics: Birds of prey with keen eyesight, hooked beaks, and talons adapted for hunting.

Examples: Eagles, hawks, and vultures.

Order Anseriformes (Waterfowl)

Characteristics: Aquatic birds with webbed feet and bills adapted for filtering food from water.

Examples: Ducks, geese, and swans.

Order Psittaciformes (Parrots)

Characteristics: Known for their bright plumage, strong beaks, and ability to mimic sounds.

Examples: Parrots, macaws, and cockatoos.

Order Strigiformes (Owls)

Characteristics: Nocturnal predators with large eyes, silent flight, and specialized hearing.

Examples: Barn owls, great horned owls.

C. Adaptations for Flight

Wing Structure

Aerodynamics: Wings are adapted for various types of flight, including gliding, hovering, and flapping.

Wing Shape: Varied according to the bird's lifestyle, such as long wings for soaring or short wings for maneuverability.

Metabolism

High Metabolic Rate: Birds have high metabolic rates to support the energy demands of flight, with a specialized digestive system for efficient food processing.

Sensory Adaptations

Vision: Many birds have excellent vision, with color perception and visual acuity adapted to their ecological needs.

2. Mammals

Mammals (Class Mammalia) are warm-blooded vertebrates characterized by the presence of hair or fur, mammary glands, and a specialized three-bone middle ear. They have diversified into various ecological roles and habitats.

A. General Characteristics

Hair

Function: Provides insulation, camouflage, and sensory input. Hair or fur varies widely among species, from dense coats in polar mammals to sparse coverings in desert-dwelling species.

Mammary Glands

Function: Females have mammary glands that produce milk to nourish their young. This feature is crucial for the survival and development of offspring.

Skeleton

Structure: Mammals possess a well-developed skeletal system, including a backbone with a specialized vertebral column. The skull structure accommodates a specialized jaw joint and middle ear bones.

Teeth

Types: Mammals have different types of teeth (incisors, canines, molars) adapted to their diet, ranging from herbivory to carnivory.

Reproduction

Modes: Mammals exhibit various reproductive strategies, including placental (eutherians), marsupial (marsupials), and monotreme (monotremes) methods.

B. Major Groups of Mammals

Monotremes

Characteristics: Egg-laying mammals with a cloaca (a single opening for excretion and reproduction).

Examples: Platypus and echidna.

Reproduction: Lay eggs that are incubated externally.

Marsupials

Characteristics: Give birth to relatively undeveloped young that continue to develop in a pouch.

Examples: Kangaroos, koalas, and wombats.

Reproduction: Young are born in a very immature state and complete development in the mother's pouch.

Eutherians (Placental Mammals)

Characteristics: Have a complex placenta that allows for a prolonged gestation period and more developed young at birth.

Examples: Humans, elephants, and bats.

Reproduction: Offspring are born at a more advanced stage of development due to the placenta providing nutrients and oxygen.

C. Adaptations for Diverse Environments

Thermoregulation

Adaptations: Mammals have various adaptations for maintaining body temperature, including sweat glands, specialized fur, and behavioral adaptations.

Dietary Specializations

Herbivores: Such as cows and deer, with specialized teeth and digestive systems for processing plant material.

Carnivores: Such as lions and wolves, with sharp teeth and claws for hunting and processing meat.

Omnivores: Such as humans and bears, with a varied diet and adaptable digestive systems.

Sensory Systems

Hearing: Mammals have highly developed auditory systems, including a three-bone middle ear that enhances sound transmission.

Smell and Vision: Many mammals have specialized olfactory and visual systems adapted to their ecological niches, such as nocturnal vision in bats or acute sense of smell in dogs.

3. Ecological and Economic Importance

A. Ecological Roles

Birds: Play crucial roles as pollinators, seed dispersers, and predators. Their presence often indicates the health of ecosystems.

Mammals: Serve various ecological functions, from predators controlling prey populations to herbivores influencing plant communities and ecosystem structure.

B. Economic Importance

Birds: Impact agriculture as pest controllers or crop protectors and contribute to ecotourism. Some species are also valued for their eggs or feathers.

Mammals: Have significant economic value in agriculture (e.g., livestock), industry (e.g., leather production), and science (e.g., research models). Many mammals are also important in tourism and recreation (e.g., wildlife watching).

Birds and mammals represent two highly evolved groups of vertebrates, each with unique adaptations that enable them to occupy diverse habitats and ecological niches. Birds, with their feathered bodies and specialized flight adaptations, and mammals, with their fur and mammary glands, showcase the diversity and complexity of vertebrate evolution. Understanding their biology, ecology, and economic importance provides valuable insights into their roles in natural ecosystems and human societies.

Part VII: Ecology and the Environment

Chapter 18: Principles of Ecology
Ecosystems and Biomes

Ecology is the study of interactions between organisms and their environments. Understanding ecosystems and biomes is crucial for comprehending how life functions on Earth and how various factors influence ecological balance and biodiversity. This chapter explores the concepts of ecosystems and biomes, examining their components, functions, and the interactions within and between them.

1. Ecosystems

An **ecosystem** is a dynamic complex of plant, animal, and microorganism communities and their non-living environment interacting as a functional unit. It encompasses both the biotic (living) and abiotic (non-living) components of an environment.

A. Components of Ecosystems

Biotic Components

Producers: Also known as autotrophs, these are organisms that produce their own food through photosynthesis (plants, algae) or chemosynthesis (certain bacteria). They form the base of the food chain.

Consumers: Organisms that consume other organisms for energy. They are categorized based on their feeding habits:

Primary Consumers: Herbivores that feed on producers (e.g., deer, insects).

Secondary Consumers: Carnivores that feed on primary consumers (e.g., snakes, birds).

Tertiary Consumers: Top predators that feed on secondary consumers (e.g., lions, eagles).

Decomposers: Organisms that break down dead organic matter, returning nutrients to the soil (e.g., bacteria, fungi, earthworms).

Abiotic Components

Climate: Includes temperature, precipitation, humidity, and light, which influence the types of organisms that can survive in an ecosystem.

Soil: Provides nutrients and a medium for plant growth. Soil composition and quality affect the types of vegetation that can thrive.

Water: Essential for all life forms. Aquatic ecosystems are influenced by factors such as water temperature, pH, salinity, and oxygen levels.

Topography: Includes features like elevation and slope that affect drainage, sunlight exposure, and soil erosion.

B. Ecosystem Processes

Energy Flow

Food Chains and Food Webs: Represent the flow of energy from producers to various levels of consumers. A food chain is a linear sequence, while a food web is a more complex network of interconnected food chains.

Trophic Levels: Each step in a food chain or web is a trophic level. Energy decreases as it moves up trophic levels due to inefficiencies in energy transfer.

Nutrient Cycling

Biogeochemical Cycles: The movement of elements and compounds (e.g., carbon, nitrogen, phosphorus) through ecosystems. Key processes include:

Carbon Cycle: Involves the movement of carbon through photosynthesis, respiration, decomposition, and combustion.

Nitrogen Cycle: Includes processes like nitrogen fixation, nitrification, assimilation, and denitrification.

Phosphorus Cycle: Involves the movement of phosphorus through the soil, water, and living organisms.

Ecological Succession

Primary Succession: Occurs in previously uninhabited areas where no soil exists (e.g., after a volcanic eruption). It starts with pioneer species like lichens and mosses.

Secondary Succession: Occurs in areas where a disturbance has cleared the ecosystem but soil remains (e.g., after a forest fire). It progresses through stages of increasing complexity until a stable climax community is reached.

2. Biomes

A **biome** is a large geographic biotic unit characterized by its climate, flora, and fauna. Biomes are distinguished by their predominant vegetation and climate, and they span multiple ecosystems.

A. Major Biomes

Tropical Rainforest

Location: Near the equator, between the Tropic of Cancer and the Tropic of Capricorn.

Climate: Warm temperatures year-round with high rainfall (2000-4000 mm annually).

Vegetation: Dense, multi-layered forests with high biodiversity. Trees have broad leaves, and there is a high level of epiphyte growth.

Fauna: Includes a vast array of species such as jaguars, sloths, monkeys, and numerous bird and insect species.

Savanna

Location: Found in tropical and subtropical regions, such as Africa, South America, and Australia.

Climate: Warm temperatures year-round with a wet and dry season. Annual rainfall ranges from 500 to 1500 mm.

Vegetation: Grasslands with scattered trees and shrubs. Vegetation is adapted to withstand periodic fires and drought.

Fauna: Includes large herbivores like elephants, giraffes, and zebras, and predators like lions and hyenas.

Desert

Location: Typically found at 30° latitude north and south, in regions like the Sahara, Arabian, and Mojave Deserts.

Climate: Extreme temperatures with very low precipitation (less than 250 mm annually). Large temperature fluctuations between day and night.

Vegetation: Sparse, with drought-resistant plants such as cacti and xerophytes.

Fauna: Includes species adapted to arid conditions, such as camels, lizards, and nocturnal mammals.

Temperate Grassland

Location: Found in central North America (prairies) and Eurasia (steppes).

Climate: Moderate temperatures with moderate to low rainfall (300-900 mm annually). Distinct seasonal variations.

Vegetation: Dominated by grasses and small shrubs. Trees are scarce due to periodic droughts and fires.

Fauna: Includes large herbivores like bison and antelope, and predators like wolves and coyotes.

Temperate Deciduous Forest

Location: Found in North America, Europe, and Asia, typically between 30° and 50° latitude.

Climate: Moderate temperatures with well-defined seasons. Annual rainfall ranges from 750 to 1500 mm.

Vegetation: Broadleaf trees that shed leaves in autumn. Includes oak, maple, and beech trees.

Fauna: Includes mammals like deer and bears, and birds like owls and woodpeckers.

Taiga (Boreal Forest)

Location: Found in high northern latitudes, such as Canada, Russia, and Scandinavia.

Climate: Cold temperatures with long winters and short, moist summers. Annual rainfall is 400-800 mm.

Vegetation: Coniferous forests with species like spruce, fir, and pine. Vegetation is adapted to cold and low-light conditions.

Fauna: Includes large herbivores like moose and caribou, and predators like lynxes and wolves.

Tundra

Location: Found in polar regions, such as Arctic Canada and Siberia.

Climate: Extremely cold temperatures with short, cool summers and long winters. Low precipitation (150-250 mm annually).

Vegetation: Low-growing plants, including mosses, lichens, and small shrubs. Permafrost limits plant growth.

Fauna: Includes animals adapted to extreme cold, such as polar bears, arctic foxes, and migratory birds.

B. Biome Characteristics and Adaptations

Adaptations

Plants: Adaptations vary widely by biome. For example, desert plants have deep roots and water storage tissues, while tropical rainforest plants have large leaves to capture more sunlight.

Animals: Adaptations include behavioral, physiological, and anatomical changes. For instance, animals in cold biomes have thick fur for insulation, while those in hot biomes may have specialized cooling mechanisms.

Human Impact

Deforestation: Particularly affects tropical rainforests and temperate forests, leading to habitat loss and biodiversity decline.

Desertification: The conversion of fertile land into desert, often due to agricultural practices and climate change.

Climate Change: Affects biomes globally, altering temperature and precipitation patterns, which can disrupt ecosystems and species distributions.

3. Interactions Between Ecosystems and Biomes

Ecotones

Definition: Transitional zones between two different ecosystems or biomes. They often have high biodiversity and unique species.

Examples: The transition between a forest and a grassland or between a lake and its surrounding land.

Biogeographical Patterns

Latitude: Affects the distribution of biomes. For example, tropical biomes are near the equator, while polar biomes are found near the poles.

Altitude: High altitudes can create alpine biomes with conditions similar to polar regions, even at lower latitudes.

Ecosystem Services

Definition: The benefits that ecosystems provide to humans, such as clean air and water, soil fertility, and climate regulation.

Examples: Forests acting as carbon sinks, wetlands filtering pollutants, and grasslands supporting agricultural productivity.

Ecosystems and biomes are fundamental concepts in ecology that illustrate the complex interactions between living organisms and their environments. Understanding these concepts helps us appreciate the diversity of life on Earth and the importance of preserving ecological balance. By studying ecosystems' structure and functions and the characteristics of different biomes, we gain insights into how natural systems operate and how human activities impact the environment.

Population Dynamics

Population dynamics is a key concept in ecology that examines the changes in population size and composition over time and the factors that influence these changes. Understanding population dynamics is crucial for managing wildlife populations, conserving endangered species, and addressing ecological and environmental issues.

1. Basic Concepts in Population Dynamics

A. Population

A **population** is a group of individuals of the same species living in a particular area and capable of interbreeding. Key characteristics of populations include size, density, distribution, age structure, and sex ratio.

B. Population Size

Definition: The number of individuals in a population.

Estimation Methods:

Direct Counts: Counting all individuals, feasible only for small or easily observable populations.

Sampling Methods: Using techniques like quadrats or transects to estimate population size in larger or less accessible areas.

Mark-Recapture Method: Individuals are captured, marked, and released. Later, a second sample is taken to estimate population size based on the proportion of marked individuals.

2. Factors Affecting Population Dynamics

A. Birth and Death Rates

Birth Rate (Natality): The rate at which new individuals are born into the population. Influenced by factors like reproductive age, fecundity (number of offspring produced), and mating behavior.

Death Rate (Mortality): The rate at which individuals die. Influenced by predation, disease, competition, and environmental conditions.

B. Immigration and Emigration

Immigration: The arrival of individuals into a population from other areas, which can increase population size.

Emigration: The departure of individuals from a population to other areas, which can decrease population size.

C. Population Growth Models

Exponential Growth Model

Description: Describes unrestricted population growth under ideal conditions with abundant resources. The population size increases rapidly over time.

Mathematical Representation: $N(t) = N_0 \times e^{(rt)}$

$N(t)$: Population size at time t

N_0: Initial population size

e: Base of the natural logarithm

r: Intrinsic rate of increase

Characteristics: J-shaped curve when graphed. This model is theoretical and rarely observed in natural populations for extended periods.

Logistic Growth Model

Description: Accounts for environmental limitations and resources. The population grows rapidly initially but slows as it approaches the carrying capacity.

Mathematical Representation: $N(t) = \frac{K}{1 + \frac{(K - N_0)}{N_0} \times e^{-rt}}$

K: Carrying capacity of the environment

The graph of this model produces an S-shaped curve (sigmoid curve), reflecting growth that stabilizes at carrying capacity.

3. Factors Influencing Population Growth

A. Density-Dependent Factors

Definition: Factors whose effects on population size vary with population density.

Examples:

Competition: Increased competition for limited resources (food, space) as population density rises.

Predation: Higher density can attract more predators, increasing mortality rates.

Disease: Higher density facilitates the spread of diseases and parasites.

B. Density-Independent Factors

Definition: Factors that affect population size regardless of population density.

Examples:

Weather: Extreme weather events (droughts, floods) can impact populations regardless of their size.

Natural Disasters: Events like volcanic eruptions or earthquakes can dramatically affect populations.

4. Life History Strategies

A. r-Selected Species

Characteristics:

High Reproductive Rates: Produce many offspring in a short period.

Short Lifespans: High mortality rates often mean individuals do not live long.

Minimal Parental Care: Offspring are often left to fend for themselves.

Examples: Insects, rodents, and annual plants.

B. K-Selected Species

Characteristics:

Low Reproductive Rates: Produce fewer offspring but invest more in their upbringing.

Long Lifespans: Individuals tend to live longer and have lower mortality rates.

Extensive Parental Care: Offspring receive significant care and protection.

Examples: Elephants, humans, and large mammals.

5. Population Interactions

A. Predation

Definition: A relationship where one organism (predator) feeds on another (prey). Influences both predator and prey populations.

Dynamics: Predator and prey populations often exhibit cyclic fluctuations, with prey populations peaking before predator populations.

B. Competition

Definition: Interaction where individuals or species vie for the same resources (food, space) that are in limited supply.

Types:

Intraspecific Competition: Competition within the same species.

Interspecific Competition: Competition between different species.

C. Mutualism

Definition: A symbiotic relationship where both species benefit.

Examples: Bees and flowering plants (pollination), clownfish and sea anemones (protection and food).

D. Parasitism

Definition: A relationship where one species (the parasite) benefits at the expense of another (the host).

Examples: Fleas on dogs, tapeworms in the intestines of animals.

6. Population Regulation

A. Carrying Capacity

Definition: The maximum population size that an environment can sustain indefinitely based on available resources.

Factors Influencing Carrying Capacity: Resource availability, habitat space, and environmental conditions.

B. Population Cycles

Definition: Regular fluctuations in population size over time, often influenced by factors like predation, competition, and environmental conditions.

Examples: Hare and lynx populations exhibit cyclic dynamics due to predator-prey interactions.

7. Human Impact on Population Dynamics

A. Habitat Destruction

Effects: Reduces available living space and resources, leading to population declines and extinctions.

B. Overexploitation

Effects: Harvesting species at unsustainable rates, leading to population declines (e.g., overfishing, hunting).

C. Pollution

Effects: Alters habitat conditions, introduces toxins, and affects reproductive success.

D. Climate Change

Effects: Alters temperature and precipitation patterns, affecting species distributions and population dynamics.

Population dynamics provides valuable insights into the processes that regulate population size and structure. By studying birth and death rates, immigration and emigration, and the factors that influence population growth, ecologists can better understand how populations interact with their environments and how they respond to changes. This knowledge is essential for effective conservation efforts, sustainable resource management, and addressing the impacts of human activities on natural ecosystems.

Chapter 19: Interactions in the Ecosystem

Symbiosis and Competition

Understanding the interactions within ecosystems is crucial for comprehending how biological communities function and maintain balance. Two fundamental types of interactions are **symbiosis** and **competition**, each playing a significant role in shaping ecological dynamics and influencing species survival and community structure.

1. Symbiosis

Symbiosis refers to a close, long-term interaction between two or more different species living in close physical proximity. This interaction can be beneficial, neutral, or harmful to the organisms involved, and it is categorized into several types:

A. Types of Symbiotic Relationships

Mutualism

Definition: A type of symbiosis where both species involved benefit from the interaction.

Examples:

Pollination: Bees and flowering plants. Bees receive nectar from flowers for food, while flowers get pollinated, facilitating reproduction.

Cleaner Fish: Cleaner fish, such as cleaner wrasses, eat parasites off larger fish (clients). The cleaner fish gain food, while the client fish get rid of parasites.

Benefits: Increases reproductive success, enhances survival, and provides essential resources like food or protection.

Commensalism

Definition: A type of symbiosis where one species benefits while the other is neither helped nor harmed.

Examples:

Birds and Large Herbivores: Birds like oxpeckers often ride on large herbivores (e.g., buffalo), eating ticks and other parasites from their skin. The herbivores are not significantly affected.

Epiphytes: Plants like orchids or bromeliads grow on trees. They obtain access to sunlight and nutrients from the air, while the host tree is neither significantly benefited nor harmed.

Benefits: Provides access to resources or habitats without affecting the host.

Parasitism

Definition: A type of symbiosis where one species benefits at the expense of the other.

Examples:

Tapeworms: Parasites that live in the intestines of their hosts, absorbing nutrients from the host's digestive system.

Fleas and Mammals: Fleas feed on the blood of mammals, causing discomfort and potential health issues for the host.

Impact: Can weaken or harm the host, potentially leading to reduced fitness or even death.

B. Importance of Symbiosis

Ecological Balance: Symbiotic relationships contribute to the stability and productivity of ecosystems by facilitating nutrient cycling, resource acquisition, and species interactions.

Evolutionary Significance: Symbiosis can drive evolutionary changes by creating selective pressures that lead to coevolution, where interacting species evolve in response to each other.

Biodiversity: Mutualistic interactions often support diverse communities by providing specialized ecological niches and enhancing species survival.

2. Competition

Competition occurs when two or more species or individuals vie for the same limited resources, such as food, water, space, or mates. This interaction can affect the growth, survival, and reproduction of the competing species.

A. Types of Competition

Intraspecific Competition

Definition: Competition within the same species.

Examples:

Territorial Animals: Male deer may compete for territory and mating rights within their population.

Plant Competition: Individual plants compete for light, water, and soil nutrients.

Effects: Can lead to density-dependent regulation of population size and influence social structures and behavior.

Interspecific Competition

Definition: Competition between different species.

Examples:

Predator-Prey Interactions: Lions and hyenas may compete for the same prey in African savannas.

Plant Competition: Different plant species may compete for sunlight and soil resources in a forest.

Effects: Can lead to niche differentiation, resource partitioning, and competitive exclusion.

B. Outcomes of Competition

Competitive Exclusion Principle

Definition: States that two species competing for the same limiting resource cannot coexist indefinitely. One species will outcompete the other, leading to the local extinction of the less competitive species or a shift to a different ecological niche.

Example: Two species of barnacles (Chthamalus and Balanus) competing for space on rocky shores. Chthamalus occupies higher zones where Balanus cannot survive.

Resource Partitioning

Definition: The process by which competing species divide resources to reduce competition and coexist.

Examples:

Birds in the Same Habitat: Different bird species may feed on different insect types or at different heights in the same tree.

Plant Root Systems: Plants may have different root depths to access various soil nutrients.

Niche Differentiation

Definition: The evolutionary process where species adapt to utilize different resources or habitats to minimize competition.

Examples:

Darwin's Finches: Beak size variation in finches on the Galápagos Islands allows different species to exploit different types of seeds.

C. Importance of Competition

Ecosystem Dynamics: Competition influences species distributions, community structure, and ecosystem functions.

Natural Selection: Drives evolutionary adaptations and speciation by creating selective pressures on competing species.

Resource Utilization: Affects how resources are used and allocated within ecosystems, impacting overall productivity and stability.

3. Interaction between Symbiosis and Competition

Coexistence and Conflict: Symbiotic relationships can sometimes mitigate competitive interactions by providing alternative resources or benefits. For instance, mutualistic relationships may help species avoid direct competition for limited resources.

Balance of Interactions: Both symbiosis and competition shape community structure and species interactions, creating a complex web of relationships that influence ecosystem health and dynamics.

4. Human Impact on Symbiosis and Competition

Habitat Destruction: Alters the availability of resources and disrupts symbiotic relationships and competitive interactions.

Invasive Species: Can outcompete native species and disrupt existing symbiotic relationships, leading to ecological imbalances.

Climate Change: Affects resource availability and environmental conditions, influencing both competitive dynamics and symbiotic interactions.

Symbiosis and competition are fundamental aspects of ecological interactions that shape the structure and dynamics of ecosystems. Understanding these interactions helps elucidate how species coexist, how ecosystems function, and how various factors influence biodiversity and ecosystem stability. By studying these relationships, ecologists can better manage and conserve natural systems, address ecological challenges, and appreciate the intricate balance of life on Earth.

Predator-Prey Relationships

Predator-prey relationships are fundamental interactions within ecosystems that drive ecological dynamics, influence community structure, and affect evolutionary processes. These relationships involve one organism (the predator) hunting and consuming another organism (the prey), which can lead to various ecological and evolutionary outcomes.

1. Characteristics of Predator-Prey Relationships

A. Predator

Definition: An organism that hunts, kills, and consumes other organisms (prey) for sustenance.

Adaptations:

Physical: Enhanced senses (e.g., sharp vision in hawks), specialized hunting tools (e.g., claws, fangs).

Behavioral: Hunting strategies (e.g., ambush, pursuit), social structures (e.g., pack hunting in wolves).

B. Prey

Definition: An organism that is hunted and consumed by predators.

Adaptations:

Physical: Camouflage (e.g., chameleons), defensive structures (e.g., quills, shells).

Behavioral: Escape behaviors (e.g., fleeing, hiding), group living (e.g., herding, schooling).

2. Dynamics of Predator-Prey Interactions

A. Population Cycles

Definition: Fluctuations in predator and prey populations over time, often exhibiting cyclical patterns.

Mechanisms:

Predator Lag: Predator populations tend to increase in response to a rise in prey populations, but with a time lag.

Prey Decline: As predator numbers increase, prey populations decline due to increased predation.

Predator Decline: A decline in prey leads to a decrease in predator populations due to lack of food resources.

Prey Recovery: Lower predator numbers allow prey populations to recover, leading to a subsequent rise in predator numbers.

B. Functional Response

Definition: The relationship between the prey density and the rate at which a predator consumes prey.

Types:

Type I: Linear increase in consumption with prey density, typical for filter feeders like baleen whales.

Type II: Decelerating consumption rate with increasing prey density, where predators reach a maximum consumption rate due to handling time (e.g., most predators).

Type III: S-shaped curve with low consumption at low prey densities, increasing at intermediate densities, and leveling off at high densities, often due to prey switching or learning (e.g., generalist predators).

C. Numerical Response

Definition: Changes in predator population size in response to changes in prey density.

Types:

Reproductive Response: Increase in predator birth rates as prey availability improves.

Immigration Response: Movement of predators into areas with high prey density.

3. Types of Predator-Prey Relationships

A. Herbivory

Definition: A form of predation where herbivores feed on plants.

Examples:

Insects and Plants: Caterpillars feeding on leaves.

Large Herbivores: Deer browsing on shrubs and grasses.

Effects: Can influence plant community structure, growth patterns, and biodiversity.

B. Carnivory

Definition: Predation involving the consumption of other animals.

Examples:

Carnivorous Mammals: Lions preying on antelopes.

Birds of Prey: Eagles hunting small mammals or birds.

Effects: Regulates prey populations, influences prey behavior, and impacts ecosystem balance.

C. Parasitism

Definition: A relationship where one organism (parasite) benefits at the expense of another (host), often without immediate death of the host.

Examples:

Internal Parasites: Tapeworms in the intestines of animals.

External Parasites: Fleas or ticks on mammals.

Effects: Can weaken hosts, reduce reproductive success, and alter host behavior.

4. Evolutionary Implications of Predator-Prey Relationships

A. Coevolution

Definition: The reciprocal evolutionary changes that occur between interacting species.

Examples:

Camouflage and Counter-Camouflage: Prey evolving to blend into their environment, while predators evolve better detection mechanisms.

Toxins and Immunities: Prey species developing chemical defenses, and predators evolving resistance or specialized detoxification mechanisms.

B. Evolutionary Arms Race

Definition: A continual cycle of adaptation and counter-adaptation between predators and prey.

Examples:

Speed and Agility: Prey species evolving faster speeds or agility to evade predators, while predators evolve improved hunting strategies.

Defensive Mechanisms: Prey evolving new defensive adaptations, prompting predators to develop new hunting techniques.

5. Ecological and Environmental Impacts

A. Community Structure

Influence on Species Diversity: Predation can maintain or enhance species diversity by preventing any single species from dominating an ecosystem.

Trophic Cascades: Changes in predator populations can impact the abundance and behavior of prey species, which in turn affects primary producers and other organisms in the food web.

B. Ecosystem Function

Nutrient Cycling: Predators contribute to nutrient cycling by affecting prey populations and influencing the decomposition process.

Habitat Structure: Herbivores can shape vegetation structure, which impacts other species and overall habitat complexity.

6. Human Impact on Predator-Prey Relationships

A. Habitat Destruction

Effects: Reduces available habitat, disrupts predator-prey interactions, and can lead to declines or extinctions of both predators and prey.

B. Overexploitation

Effects: Excessive hunting or fishing can deplete prey populations, which in turn affects predator populations and overall ecosystem health.

C. Introduced Species

Effects: Non-native predators or prey can disrupt existing predator-prey relationships, leading to ecological imbalances and potentially the decline of native species.

D. Climate Change

Effects: Alters habitat conditions, affects prey availability, and can shift predator-prey dynamics by changing the timing of biological events (e.g., breeding, migration).

Predator-prey relationships are integral to ecosystem dynamics and evolutionary processes. They influence population sizes, community structure, and species interactions. By studying these relationships, ecologists can better understand the complexities of natural systems, manage wildlife populations, and address the impacts of human activities on ecosystems. The interplay between predators and prey exemplifies the delicate balance of nature and highlights the intricate web of life that sustains ecological harmony.

Human Impact on Ecosystems

Human activities have a profound impact on ecosystems, influencing their structure, function, and biodiversity. Understanding these impacts is crucial for developing strategies to mitigate damage, conserve natural resources, and sustain ecological balance. This chapter explores the various ways in which human actions affect ecosystems and discusses potential solutions to address these issues.

1. Habitat Destruction

A. Deforestation

Definition: The large-scale removal of forests for agriculture, urban development, or logging.

Impacts:

Loss of Biodiversity: Deforestation leads to the loss of habitats for many species, causing declines in wildlife populations and extinctions.

Climate Change: Trees act as carbon sinks; their removal increases atmospheric CO_2 levels, contributing to global warming.

Disruption of Water Cycles: Forests play a role in maintaining local and global water cycles. Their removal can alter precipitation patterns and lead to reduced water availability.

B. Urbanization

Definition: The expansion of cities and towns into natural habitats.

Impacts:

Fragmentation: Urban sprawl fragments habitats, creating isolated patches that can be difficult for wildlife to navigate and survive in.

Pollution: Urban areas often increase pollution levels (air, water, noise), which can harm both wildlife and human health.

Altered Ecosystem Services: Urbanization reduces the ability of natural systems to provide ecosystem services such as water filtration, climate regulation, and pollination.

C. Agricultural Expansion

Definition: The conversion of natural landscapes into farmland.

Impacts:

Loss of Natural Habitats: Conversion of forests, grasslands, and wetlands into cropland reduces biodiversity and disrupts ecosystems.

Soil Degradation: Intensive farming practices can lead to soil erosion, loss of soil fertility, and desertification.

Chemical Pollution: Use of fertilizers and pesticides can contaminate soil and water, harming non-target species and ecosystems.

2. Pollution

A. Air Pollution

Sources: Emissions from vehicles, industries, and burning fossil fuels.

Impacts:

Acid Rain: Airborne pollutants such as sulfur dioxide and nitrogen oxides can cause acid rain, which harms plant life, soil, and water bodies.

Climate Change: Greenhouse gases contribute to global warming, affecting climate patterns and ecosystems.

Health Effects: Air pollution can cause respiratory problems and other health issues in humans and wildlife.

B. Water Pollution

Sources: Industrial discharge, agricultural runoff, sewage, and plastic waste.

Impacts:

Eutrophication: Excess nutrients from fertilizers can lead to algal blooms in water bodies, depleting oxygen and harming aquatic life.

Toxic Contaminants: Heavy metals, pesticides, and other pollutants can poison aquatic organisms and disrupt food chains.

Habitat Destruction: Pollutants can degrade or destroy aquatic habitats such as coral reefs and wetlands.

C. Soil Pollution

Sources: Pesticides, heavy metals, and industrial waste.

Impacts:

Decreased Soil Fertility: Contaminants can reduce the ability of soil to support plant growth.

Health Risks: Polluted soil can affect human health through the consumption of contaminated crops or direct exposure.

Ecosystem Disruption: Soil pollution can impact soil-dwelling organisms, leading to reduced decomposition and nutrient cycling.

3. Climate Change

A. Greenhouse Gas Emissions

Sources: Burning of fossil fuels, deforestation, and industrial processes.

Impacts:

Temperature Rise: Increased greenhouse gases lead to global warming, affecting temperature-sensitive ecosystems and species.

Ocean Acidification: Increased CO2 levels cause ocean acidification, harming marine life, particularly organisms with calcium carbonate shells or skeletons.

Sea Level Rise: Melting polar ice and thermal expansion contribute to rising sea levels, threatening coastal habitats and communities.

B. Altered Weather Patterns

Impacts:

Increased Frequency of Extreme Weather: More frequent and severe weather events such as hurricanes, droughts, and floods can damage ecosystems and human infrastructure.

Shifts in Species Distribution: Changes in temperature and precipitation patterns can cause species to migrate or adapt, disrupting existing ecosystems and community structures.

4. Overexploitation

A. Overfishing

Definition: Harvesting fish and other marine organisms at rates that exceed their ability to replenish.

Impacts:

Depletion of Fish Stocks: Overfishing can lead to declines in fish populations, affecting marine food webs and economies.

Bycatch: Non-target species are often caught unintentionally, leading to declines in marine biodiversity.

Ecosystem Disruption: Removing key species can alter marine ecosystems and affect the balance of predator-prey relationships.

B. Hunting and Poaching

Definition: Illegal or unsustainable hunting of wildlife for meat, trophies, or traditional medicine.

Impacts:

Population Declines: Overhunting can lead to population declines and extinction of targeted species.

Disruption of Ecosystems: Removing top predators or key species can disrupt ecological balance and food chains.

Loss of Biodiversity: Poaching and hunting reduce species diversity and ecosystem resilience.

C. Resource Extraction

Definition: Mining, drilling, and other activities that extract natural resources.

Impacts:

Habitat Destruction: Resource extraction often involves clearing land and disturbing ecosystems.

Pollution: Mining and drilling can release pollutants into air, water, and soil, causing environmental damage.

Biodiversity Loss: Disruption of habitats and introduction of pollutants can lead to declines in species populations and loss of biodiversity.

5. Invasive Species

A. Introduction of Non-Native Species

Definition: The introduction of species to new environments where they do not naturally occur.

Impacts:

Competition: Invasive species can outcompete native species for resources, leading to declines or extinctions of native species.

Predation: Some invasive species prey on native species, disrupting local food webs.

Alteration of Ecosystems: Invasive species can change habitat structure, nutrient cycling, and other ecological processes.

6. Conservation and Mitigation Strategies

A. Protected Areas

Definition: Designated regions where human activities are restricted to protect biodiversity and natural habitats.

Types: National parks, wildlife reserves, marine protected areas.

Benefits: Helps preserve critical habitats, protect endangered species, and maintain ecosystem functions.

B. Sustainable Practices

Definition: Practices that aim to meet current needs without compromising the ability of future generations to meet their needs.

Examples:

Sustainable Agriculture: Practices that reduce environmental impact, such as crop rotation, organic farming, and reduced use of chemicals.

Sustainable Fishing: Methods that prevent overfishing and protect marine ecosystems, such as quotas and selective fishing gear.

C. Restoration Ecology

Definition: The science of restoring and rehabilitating degraded ecosystems to their natural conditions.

Methods: Reforestation, wetland restoration, removal of invasive species.

Benefits: Aims to recover ecosystem functions, improve biodiversity, and enhance ecosystem resilience.

D. Environmental Policies and Legislation

Examples:

Endangered Species Act: Protects endangered and threatened species and their habitats.

Clean Water Act: Regulates pollutant discharges into water bodies and ensures water quality.

Paris Agreement: An international treaty aimed at limiting global temperature rise and addressing climate change.

Human activities have significant and far-reaching impacts on ecosystems, affecting their structure, function, and biodiversity. Addressing these impacts requires a comprehensive approach that includes habitat protection, sustainable practices, restoration efforts, and effective environmental policies. By understanding and mitigating the effects of human actions on ecosystems, we can work towards maintaining ecological balance, conserving natural resources, and ensuring a sustainable future for all species, including humans.

Conservation Strategies

Conservation biology is a scientific discipline focused on understanding and mitigating the loss of biodiversity. Effective conservation strategies aim to protect species, habitats, and ecosystems from degradation, ensuring that they remain viable and resilient for future generations. This chapter delves into the various conservation strategies employed to address the challenges of biodiversity loss and ecosystem degradation.

1. Protected Areas

A. Types of Protected Areas

National Parks

Purpose: Protect large natural landscapes and biodiversity, providing opportunities for recreation and education.

Examples: Yellowstone National Park (USA), Serengeti National Park (Tanzania).

Wildlife Reserves

Purpose: Safeguard specific species or habitats, often with restrictions on human activities to minimize disturbance.

Examples: Kruger National Park (South Africa), Great Barrier Reef Marine Park (Australia).

Marine Protected Areas (MPAs)

Purpose: Protect marine ecosystems and species from overexploitation, pollution, and habitat destruction.

Examples: Papahānaumokuākea Marine National Monument (Hawaii), Galápagos Marine Reserve (Ecuador).

Biosphere Reserves

Purpose: Integrate conservation with sustainable development, involving local communities in management.

Examples: Mura-Drava-Danube Biosphere Reserve (Europe), Sundarbans Biosphere Reserve (India/Bangladesh).

B. Benefits and Challenges

Benefits:

Biodiversity Protection: Preserves habitats and species, aiding in the recovery of endangered species.

Ecosystem Services: Maintains ecosystem functions such as water purification, carbon sequestration, and soil fertility.

Education and Research: Provides opportunities for scientific research and environmental education.

Challenges:

Insufficient Coverage: Many protected areas are too small or poorly situated to effectively protect biodiversity.

Management Issues: Limited resources, insufficient enforcement, and conflicts with local communities can undermine effectiveness.

Climate Change: Protected areas may not be able to accommodate species migrations in response to climate shifts.

2. Habitat Restoration

A. Goals of Restoration

Ecosystem Recovery: Restore degraded habitats to their natural state, improving ecological health and biodiversity.

Species Reintroduction: Reintroduce species to areas where they have been extirpated, restoring ecological balance.

Functional Integrity: Enhance ecosystem functions such as nutrient cycling, water regulation, and soil stabilization.

B. Methods of Restoration

Reforestation and Afforestation

Reforestation: Planting trees in deforested areas to restore forest ecosystems.

Afforestation: Establishing forests in previously non-forested areas.

Wetland Restoration

Methods: Rehabilitate wetlands by removing drainage systems, restoring natural water flow, and replanting native vegetation.

Grassland Restoration

Methods: Reintroduce native grasses and remove invasive species to restore grassland ecosystems.

Coral Reef Restoration

Methods: Transplanting corals, reducing pollution, and establishing marine protected areas to rehabilitate coral reefs.

C. Benefits and Challenges

Benefits:

Biodiversity Enhancement: Supports the recovery of plant and animal species, increasing ecosystem diversity.

Ecosystem Function: Restores critical ecosystem processes and services.

Challenges:

Cost and Time: Restoration projects can be expensive and time-consuming, requiring long-term commitment and monitoring.

Complexity: Ecosystems are complex, and restoration may not always fully replicate natural conditions.

Invasive Species: Managing invasive species can be challenging during the restoration process.

3. Sustainable Practices

A. Sustainable Agriculture

Practices:

Crop Rotation: Alternating crops to maintain soil fertility and reduce pest infestations.

Organic Farming: Avoiding synthetic chemicals and using natural methods to manage pests and soil health.

Agroforestry: Integrating trees and shrubs into agricultural landscapes to enhance biodiversity and ecosystem services.

B. Sustainable Fisheries

Practices:

Catch Limits: Implementing quotas to prevent overfishing and ensure fish populations remain sustainable.

Bycatch Reduction: Using selective fishing gear to minimize the capture of non-target species.

Marine Spatial Planning: Designating fishing zones and protecting critical habitats to balance conservation and resource use.

C. Sustainable Forestry

Practices:

Selective Logging: Harvesting specific trees while preserving the forest structure and biodiversity.

Certification Programs: Using standards such as those from the Forest Stewardship Council (FSC) to ensure sustainable forest management.

D. Benefits and Challenges

Benefits:

Resource Conservation: Ensures that natural resources are used efficiently and remain available for future generations.

Ecosystem Health: Maintains ecosystem integrity and services, reducing environmental impact.

Challenges:

Implementation: Transitioning to sustainable practices can be challenging and requires commitment from all stakeholders.

Economic Pressure: Economic incentives may favor unsustainable practices, especially in developing regions.

4. Species Conservation

A. Protected Species

Strategies:

Legal Protection: Enforcing laws and regulations to protect endangered and threatened species from exploitation.

Captive Breeding: Breeding species in captivity to increase population numbers and support reintroduction efforts.

B. Conservation Genetics

Purpose: Using genetic information to support conservation efforts, such as managing genetic diversity and understanding species relationships.

Methods:

Genetic Monitoring: Tracking genetic variation within populations to inform management decisions.

Population Viability Analysis: Assessing the likelihood of species survival based on genetic data and demographic factors.

C. Benefits and Challenges

Benefits:

Species Recovery: Targeted efforts can lead to the recovery of endangered species and prevent extinctions.

Genetic Health: Maintaining genetic diversity enhances species resilience to environmental changes.

Challenges:

Resource Intensiveness: Species conservation can require significant financial and human resources.

Conflicts: Balancing conservation goals with human interests, such as land use and economic development, can be challenging.

5. Environmental Education and Advocacy

A. Education Programs

Purpose: Raise awareness about conservation issues and promote sustainable practices.

Methods:

School Programs: Integrating conservation topics into school curricula to educate students about biodiversity and environmental stewardship.

Community Outreach: Engaging local communities through workshops, seminars, and public campaigns.

B. Advocacy and Policy

Purpose: Influence policy decisions and promote conservation efforts at local, national, and international levels.

Methods:

Lobbying: Advocating for environmental policies and legislation to protect ecosystems and species.

Public Campaigns: Using media and public events to raise awareness and support for conservation issues.

C. Benefits and Challenges

Benefits:

Increased Awareness: Education and advocacy can lead to greater public support for conservation initiatives.

Policy Change: Effective advocacy can result in stronger environmental policies and regulations.

Challenges:

Resistance: Overcoming resistance from stakeholders with conflicting interests can be difficult.

Funding: Securing adequate funding for education and advocacy programs can be challenging.

6. Global Conservation Initiatives

A. International Agreements

Examples:

Convention on Biological Diversity (CBD): An international treaty aimed at conserving biodiversity, promoting sustainable use, and ensuring fair sharing of benefits.

CITES (Convention on International Trade in Endangered Species): Regulates international trade of endangered species to prevent exploitation.

B. Global Networks and Partnerships

Examples:

World Wildlife Fund (WWF): An international organization focused on conservation and reducing human impact on the environment.

International Union for Conservation of Nature (IUCN): Provides scientific knowledge and advocacy for conservation.

C. Benefits and Challenges

Benefits:

Collaboration: Global initiatives foster collaboration among countries, organizations, and stakeholders.

Unified Goals: International agreements provide a framework for addressing global conservation challenges.

Challenges:

Implementation: Ensuring effective implementation of international agreements can be difficult due to varying national priorities and capacities.

Coordination: Coordinating efforts across different countries and organizations requires strong communication and cooperation.

Conservation strategies are essential for addressing the myriad challenges faced by ecosystems and biodiversity. By implementing effective measures such as establishing protected areas, restoring habitats, adopting sustainable practices, and supporting species conservation, we can work towards preserving the natural world for future generations. Collaboration among governments, organizations, communities, and individuals is crucial for the success of these strategies, as well as for fostering a global culture of environmental stewardship and respect for the interconnected web of life.

Sustainable Practices

Sustainable practices are fundamental to conservation biology, aiming to manage natural resources and ecosystems in a way that meets current needs while preserving the environment for future generations. These practices are designed to minimize environmental impact, promote resource conservation, and enhance ecosystem health. This chapter explores various sustainable practices across different sectors, their benefits, and the challenges associated with their implementation.

1. Sustainable Agriculture

A. Principles of Sustainable Agriculture

Resource Efficiency: Utilizing resources such as water, soil, and nutrients more efficiently to minimize waste and environmental impact.

Ecosystem Health: Maintaining soil fertility, biodiversity, and ecosystem services to support long-term agricultural productivity.

Economic Viability: Ensuring that farming practices are economically viable for farmers while minimizing environmental costs.

B. Sustainable Agricultural Practices

Crop Rotation

Description: Alternating the types of crops grown on a particular piece of land each season.

Benefits:

Soil Health: Reduces soil erosion and nutrient depletion, improves soil structure and fertility.

Pest and Disease Management: Interrupts pest and disease cycles, reducing the need for chemical pesticides.

Organic Farming

Description: Using natural methods to manage pests and soil health, avoiding synthetic chemicals and genetically modified organisms (GMOs).

Benefits:

Biodiversity: Enhances soil and crop biodiversity, improves ecosystem resilience.

Soil Health: Increases organic matter and improves soil structure and fertility.

Agroforestry

Description: Integrating trees and shrubs into agricultural landscapes.

Benefits:

Soil Conservation: Reduces soil erosion and enhances soil fertility.

Biodiversity: Provides habitat for wildlife and increases plant diversity.

Conservation Tillage

Description: Minimizing soil disturbance through reduced or no-till farming practices.

Benefits:

Soil Health: Reduces soil erosion, improves water retention, and enhances soil organic matter.

Integrated Pest Management (IPM)

Description: Combining biological, cultural, physical, and chemical methods to manage pests in an environmentally and economically sustainable manner.

Benefits:

Pest Control: Reduces reliance on chemical pesticides, minimizes environmental impact.

C. Benefits and Challenges

Benefits:

Environmental Protection: Reduces pollution and degradation of natural resources.

Economic Efficiency: Can lead to cost savings and improved profitability in the long term.

Social Benefits: Promotes healthier food systems and can improve community well-being.

Challenges:

Transition Costs: Initial costs and learning curves for adopting new practices.

Market Demand: Limited market access and consumer demand for sustainably produced goods.

Knowledge and Training: Need for education and support for farmers to implement sustainable practices effectively.

2. Sustainable Fisheries

A. Principles of Sustainable Fisheries

Resource Management: Ensuring fish stocks are harvested at rates that do not compromise their ability to replenish.

Ecosystem Health: Protecting marine ecosystems and non-target species from the impacts of fishing.

Economic and Social Benefits: Supporting livelihoods and economies dependent on fishing while preserving fishery resources.

B. Sustainable Fishing Practices

Catch Limits and Quotas

Description: Setting limits on the number of fish that can be harvested to prevent overfishing.

Benefits:

Fish Stock Recovery: Allows fish populations to rebuild and maintain healthy levels.

Ecosystem Balance: Helps sustain marine food webs and ecosystem functions.

Bycatch Reduction

Description: Implementing measures to reduce the capture of non-target species.

Methods:

Selective Fishing Gear: Using gear designed to target specific species and avoid bycatch.

Bycatch Reduction Devices: Installing devices on fishing gear to reduce bycatch of vulnerable species.

Benefits:

Biodiversity Protection: Reduces the impact on non-target species and protects marine biodiversity.

Marine Spatial Planning

Description: Designating specific areas for different types of marine activities to balance conservation and resource use.

Benefits:

Habitat Protection: Protects critical habitats from destructive activities.

Conflict Management: Reduces conflicts between different marine uses (e.g., fishing, tourism, conservation).

Community-Based Management

Description: Involving local communities in the management and conservation of fisheries.

Benefits:

Local Knowledge: Utilizes local knowledge and practices for effective management.

Stakeholder Engagement: Promotes cooperation and compliance among fishers.

C. Benefits and Challenges

Benefits:

Sustainable Resources: Ensures long-term viability of fish stocks and marine ecosystems.

Economic Stability: Supports fishing communities by maintaining fishery resources.

Challenges:

Enforcement: Monitoring and enforcing regulations can be challenging, especially in remote areas.

Economic Pressure: Short-term economic pressures may lead to non-compliance with sustainable practices.

Data Limitations: Inadequate data on fish stocks and ecosystems can hinder effective management.

3. Sustainable Forestry

A. Principles of Sustainable Forestry

Resource Management: Harvesting timber and non-timber products in a way that maintains forest health and productivity.

Biodiversity Conservation: Protecting forest ecosystems and species diversity.

Economic and Social Benefits: Supporting forest-dependent communities and economies while conserving forest resources.

B. Sustainable Forestry Practices

Selective Logging

Description: Harvesting specific trees while preserving the overall forest structure.

Benefits:

Forest Health: Reduces habitat destruction and maintains ecosystem functions.

Biodiversity: Minimizes impact on wildlife and plant species.

Reduced Impact Logging (RIL)

Description: Techniques designed to minimize environmental damage during logging operations.

Methods:

Planning and Mapping: Carefully planning logging routes and locations to avoid sensitive areas.

Minimized Soil Disturbance: Using methods that reduce soil erosion and compaction.

Benefits:

Ecosystem Protection: Reduces damage to forest soils, waterways, and non-target species.

Certification Programs

Description: Programs that certify forests and products as sustainably managed.

Examples:

Forest Stewardship Council (FSC): Provides certification for responsibly managed forests.

Program for the Endorsement of Forest Certification (PEFC): Certifies sustainable forest management practices.

Benefits:

Consumer Trust: Provides assurance to consumers that products come from sustainably managed sources.

Market Access: Opens access to markets that require certified products.

C. Benefits and Challenges

Benefits:

Forest Conservation: Preserves forest ecosystems and supports biodiversity.

Sustainable Livelihoods: Provides long-term economic benefits to communities dependent on forests.

Challenges:

Cost: Implementing and maintaining sustainable practices can be expensive.

Compliance: Ensuring compliance with sustainable practices and certification standards can be challenging.

Deforestation Pressures: Economic pressures and illegal logging can undermine conservation efforts.

4. Sustainable Resource Management

A. Principles of Sustainable Resource Management

Resource Efficiency: Using resources such as water, minerals, and energy in ways that minimize waste and environmental impact.

Ecosystem Integrity: Protecting ecosystems and their services while managing resource extraction.

Economic and Social Equity: Ensuring fair access to resources and benefits for all stakeholders.

B. Sustainable Resource Management Practices

Water Management

Description: Managing water resources to ensure availability and quality.

Practices:

Water Conservation: Implementing measures to reduce water use and waste.

Pollution Control: Reducing contaminants entering water bodies through treatment and best practices.

Benefits:

Ecosystem Health: Maintains healthy aquatic ecosystems and supports biodiversity.

Resource Availability: Ensures water availability for future needs.

Mineral Resource Management

Description: Managing the extraction and use of mineral resources to minimize environmental impact.

Practices:

Recycling: Reusing minerals from products and reducing the need for new extraction.

Environmental Monitoring: Monitoring and mitigating the environmental impacts of mining operations.

Benefits:

Resource Conservation: Reduces the need for new mining and decreases environmental damage.

Waste Reduction: Minimizes waste and pollution associated with mineral extraction.

Energy Management

Description: Managing energy use to reduce consumption and environmental impact.

Practices:

Renewable Energy: Utilizing renewable energy sources such as solar, wind, and hydro power.

Energy Efficiency: Improving energy efficiency in buildings, transportation, and industrial processes.

Benefits:

Climate Mitigation: Reduces greenhouse gas emissions and reliance on fossil fuels.

Sustainable Development: Supports long-term energy security and reduces environmental impact.

C. Benefits and Challenges

Benefits:

Resource Preservation: Ensures that resources are available for future generations.

Environmental Protection: Reduces environmental impact and supports ecosystem health.

Economic Efficiency: Promotes efficient use of resources and reduces waste.

Challenges:

Implementation: Requires significant investment and changes in practices.

Market Pressures: Economic pressures may lead to unsustainable resource use.

Coordination: Effective management requires coordination among various stakeholders and sectors.

5. Sustainable Urban Development

A. Principles of Sustainable Urban Development

Resource Efficiency: Designing cities and infrastructure to minimize resource use and environmental impact.

Ecosystem Integration: Incorporating natural elements and ecosystem services into urban planning.

Social Equity: Ensuring that urban development benefits all residents and supports equitable access to resources and services.

B. Sustainable Urban Development Practices

Green Building Design

Description: Designing and constructing buildings that are energy-efficient, environmentally friendly, and resource-conserving.

Practices:

Energy Efficiency: Using energy-efficient technologies and materials.

Sustainable Materials: Choosing materials that have a low environmental impact.

Benefits:

Reduced Environmental Impact: Lowers energy consumption and resource use.

Improved Health: Creates healthier living environments with better air quality.

Urban Green Spaces

Description: Integrating parks, gardens, and natural areas into urban environments.

Benefits:

Biodiversity: Provides habitat for wildlife and supports urban biodiversity.

Human Well-being: Enhances quality of life, mental health, and social cohesion.

Sustainable Transportation

Description: Promoting transportation systems that reduce environmental impact and support sustainable development.

Practices:

Public Transit: Expanding and improving public transportation options.

Active Transportation: Encouraging walking and cycling through infrastructure and planning.

Benefits:

Reduced Emissions: Lowers greenhouse gas emissions and air pollution.

Enhanced Mobility: Improves access and reduces traffic congestion.

C. Benefits and Challenges

Benefits:

Environmental Protection: Reduces urban environmental impact and enhances ecosystem services.

Social Benefits: Improves quality of life and promotes social equity.

Challenges:

Cost: Implementing sustainable practices can be expensive.

Infrastructure Limitations: Requires changes to existing infrastructure and planning systems.

Resistance to Change: Overcoming resistance from stakeholders with competing interests.

Sustainable practices are essential for the conservation of natural resources and ecosystems, providing a pathway to balance human needs with environmental protection. From agriculture and fisheries to forestry and urban development, adopting sustainable practices helps to preserve biodiversity, maintain ecosystem health, and support the well-being of current and future generations. However, the successful implementation of these practices requires overcoming various challenges, including economic pressures, resource limitations, and resistance to change. By fostering collaboration, investing in innovative solutions, and promoting awareness, society can work towards a more sustainable and resilient future.

Homeostasis and Regulation

Homeostasis is the process by which living organisms maintain a stable internal environment despite changes in external conditions. This equilibrium is crucial for the proper functioning of cells, tissues, and organs, enabling the body to operate efficiently and survive in various conditions. **Regulation** refers to the mechanisms and systems that the body employs to achieve and maintain homeostasis.

1. Principles of Homeostasis

A. Definition and Importance

Definition: Homeostasis is the maintenance of a stable internal environment within the body, including factors such as temperature, pH, glucose levels, and fluid balance.

Importance:

Cell Function: Cells function optimally within a narrow range of conditions. Deviations can impair cellular processes or cause cell damage.

Organ Function: Organs and systems rely on a stable internal environment to perform their specialized functions.

Overall Health: Maintaining homeostasis is essential for overall health and well-being. Disruptions can lead to diseases or disorders.

B. Homeostatic Control Mechanisms

Receptors

Function: Detect changes in the internal or external environment (stimuli).

Types:

Thermoreceptors: Monitor temperature changes.

Chemoreceptors: Detect chemical changes, such as pH or glucose levels.

Baroreceptors: Sense changes in blood pressure.

Control Center

Function: Processes the information received from receptors and determines the appropriate response.

Examples:

Hypothalamus: Acts as the control center for temperature regulation and other homeostatic functions.

Pancreas: Regulates blood glucose levels by releasing insulin or glucagon.

Effectors

Function: Carry out the response to restore homeostasis.

Types:

Muscles: E.g., shivering or sweating to regulate temperature.

Glands: E.g., sweat glands or salivary glands that adjust secretion rates based on feedback.

Feedback Mechanisms

Positive Feedback: Amplifies the response to a stimulus, moving the system away from equilibrium (e.g., childbirth contractions).

Negative Feedback: Counteracts deviations from the set point to restore equilibrium (e.g., temperature regulation).

2. Temperature Regulation

A. Mechanisms for Temperature Regulation

Thermoreceptors

Location: Found in the skin and hypothalamus.

Function: Detect changes in external and internal temperatures.

Hypothalamus

Role: Acts as the primary control center for temperature regulation.

Response: Initiates responses to adjust body temperature based on input from thermoreceptors.

Effectors

Sweat Glands: Produce sweat to cool the body through evaporation.

Blood Vessels: Dilate (vasodilation) to release heat or constrict (vasoconstriction) to retain heat.

Muscles: Shivering generates heat through muscle contractions.

B. Examples of Temperature Regulation

Heat Stress

Response: Sweating, increased heart rate, and vasodilation to dissipate heat.

Challenge: Risk of dehydration and heat exhaustion if mechanisms are overwhelmed.

Cold Stress

Response: Shivering, reduced blood flow to the skin, and piloerection (hair standing up) to conserve heat.

Challenge: Risk of hypothermia if body heat loss exceeds production.

3. Blood Glucose Regulation

A. Role of the Pancreas

Insulin

Function: Lowers blood glucose levels by promoting glucose uptake into cells and stimulating glycogen synthesis.

Produced by: Beta cells of the pancreatic islets.

Glucagon

Function: Raises blood glucose levels by stimulating glycogen breakdown and glucose release from the liver.

Produced by: Alpha cells of the pancreatic islets.

B. Regulation Mechanisms

High Blood Glucose

Stimulus: Elevated glucose levels after a meal.

Response: Insulin is released, facilitating glucose uptake by cells and glycogen formation in the liver.

Low Blood Glucose

Stimulus: Decreased glucose levels between meals or during exercise.

Response: Glucagon is released, promoting glycogen breakdown and glucose release into the bloodstream.

C. Disorders

Diabetes Mellitus

Type 1 Diabetes: Autoimmune destruction of insulin-producing beta cells.

Type 2 Diabetes: Insulin resistance and/or impaired insulin secretion.

Management: Involves monitoring blood glucose levels, dietary changes, and medication or insulin therapy.

4. Fluid and Electrolyte Balance

A. Mechanisms of Fluid Regulation

Kidneys

Function: Regulate fluid balance by adjusting the volume and composition of urine.

Processes: Filtration, reabsorption, and secretion of water and electrolytes.

Hormones

Antidiuretic Hormone (ADH): Increases water reabsorption in the kidneys, reducing urine output.

Aldosterone: Promotes sodium reabsorption and potassium excretion in the kidneys, affecting fluid balance.

B. Electrolyte Balance

Sodium and Potassium

Role: Maintain cellular function, nerve transmission, and muscle contraction.

Regulation: Controlled by hormones like aldosterone and the renal system.

Calcium

Role: Important for bone health, muscle function, and nerve signaling.

Regulation: Controlled by parathyroid hormone (PTH) and calcitonin.

C. Disorders

Dehydration

Cause: Excessive loss of fluids and electrolytes, leading to decreased blood volume and impaired function.

Symptoms: Thirst, dry skin, decreased urine output, and dizziness.

Electrolyte Imbalance

Cause: Disruption in the balance of electrolytes due to conditions like kidney disease or excessive sweating.

Symptoms: Muscle cramps, irregular heartbeat, and confusion.

5. Acid-Base Balance

A. pH Regulation

Blood pH

Normal Range: 7.35 - 7.45.

Regulation: Maintained through buffer systems, respiratory adjustments, and renal function.

Buffer Systems

Bicarbonate-Carbonic Acid Buffer: Neutralizes excess acids or bases in the blood.

Protein Buffers: Proteins in blood act as buffers by binding or releasing hydrogen ions.

B. Respiratory and Renal Regulation

Respiratory System

Function: Regulates blood pH by adjusting the rate of CO_2 exhalation.

Response: Increased breathing rate lowers CO_2 levels, reducing acidity.

Renal System

Function: Excretes excess acids or bases through urine.

Response: Adjusts the concentration of hydrogen ions and bicarbonate in the urine.

C. Disorders

Acidosis

Types: Metabolic acidosis (due to increased acid production or decreased bicarbonate) and respiratory acidosis (due to CO2 retention).

Symptoms: Fatigue, confusion, and shortness of breath.

Alkalosis

Types: Metabolic alkalosis (due to excessive bicarbonate) and respiratory alkalosis (due to excessive CO2 loss).

Symptoms: Muscle twitching, irritability, and nausea.

6. Feedback Systems

A. Negative Feedback

Description: A system where a change in a variable triggers a response that counteracts the initial change.

Example: Temperature regulation (body heat increases → sweating occurs to cool down).

B. Positive Feedback

Description: A system where a change in a variable triggers a response that amplifies the initial change.

Example: Labor contractions during childbirth (contractions increase → more oxytocin is released, leading to stronger contractions).

Homeostasis and regulation are vital processes that ensure the human body's internal environment remains stable and conducive to life. Through complex mechanisms involving receptors, control centers, effectors, and feedback systems, the body maintains equilibrium across various parameters such as temperature, blood glucose levels, fluid balance, and pH. Understanding these processes is crucial for diagnosing and managing health conditions and appreciating the intricate balance necessary for sustaining life.

Chapter 22: Nutrition and Digestion

Nutrients and Their Functions

Nutrients are substances that organisms need to live and grow. They are essential for various physiological processes, including energy production, tissue repair, and maintenance of metabolic functions. Proper nutrition is crucial for maintaining health, supporting growth, and preventing diseases. This chapter will extensively discuss the types of nutrients, their functions, sources, and the role they play in human health.

1. Macronutrients

Macronutrients are nutrients required in large amounts that provide energy and are essential for growth and maintenance. They include carbohydrates, proteins, and fats.

A. Carbohydrates

Functions

Energy Production: Carbohydrates are the primary source of energy for the body. They are broken down into glucose, which is used by cells to produce ATP (adenosine triphosphate), the energy currency of the cell.

Protein Sparing: Carbohydrates help prevent the use of proteins for energy, allowing proteins to be used for tissue repair and other functions.

Digestive Health: Dietary fiber, a type of carbohydrate, aids in digestion by adding bulk to stool and promoting regular bowel movements.

Types

Simple Carbohydrates: Include sugars such as glucose, fructose, and sucrose. Found in fruits, honey, and refined sugar products.

Complex Carbohydrates: Include starches and fibers. Found in grains, legumes, vegetables, and fruits.

Sources

Grains: Whole grains like brown rice, oats, and whole wheat.

Vegetables: Leafy greens, potatoes, and corn.

Fruits: Apples, bananas, and berries.

Legumes: Beans, lentils, and peas.

B. Proteins

Functions

Structural Role: Proteins are essential for building and repairing tissues. They are integral components of muscles, skin, hair, and nails.

Enzyme Function: Many enzymes are proteins that catalyze biochemical reactions in the body.

Hormonal Role: Some hormones, like insulin and growth hormone, are proteins that regulate physiological processes.

Immune Function: Antibodies, which help defend against infections, are proteins.

Amino Acids

Essential Amino Acids: Must be obtained from the diet as the body cannot synthesize them. Includes histidine, isoleucine, leucine, lysine, methionine, phenylalanine, threonine, tryptophan, and valine.

Non-Essential Amino Acids: Can be synthesized by the body. Includes alanine, asparagine, aspartic acid, and glutamic acid.

Sources

Animal Products: Meat, poultry, fish, eggs, and dairy products.

Plant-Based Sources: Beans, lentils, tofu, quinoa, and nuts.

C. Fats

Functions

Energy Storage: Fats provide a concentrated source of energy and are stored in adipose tissue for later use.

Cell Structure: Essential components of cell membranes, contributing to membrane fluidity and integrity.

Absorption of Vitamins: Facilitate the absorption of fat-soluble vitamins (A, D, E, and K).

Protection and Insulation: Provide cushioning for vital organs and insulation to maintain body temperature.

Types

Saturated Fats: Typically solid at room temperature. Found in animal products and some plant oils like coconut oil.

Unsaturated Fats: Usually liquid at room temperature. Include monounsaturated and polyunsaturated fats. Found in olive oil, nuts, seeds, and fish.

Trans Fats: Artificially created fats found in some processed foods. Known to increase the risk of cardiovascular diseases.

Sources

Healthy Sources: Avocados, nuts, seeds, olive oil, and fatty fish like salmon.

Less Healthy Sources: Red meat, butter, and processed snack foods.

2. Micronutrients

Micronutrients are required in smaller amounts but are essential for various bodily functions. They include vitamins and minerals.

A. Vitamins

Water-Soluble Vitamins

Vitamin C: Important for collagen synthesis, immune function, and antioxidant protection. Found in citrus fruits, strawberries, and bell peppers.

B Vitamins: Include B1 (thiamine), B2 (riboflavin), B3 (niacin), B6 (pyridoxine), B12 (cobalamin), and folate. Essential for energy metabolism, red blood cell formation, and neurological function. Found in whole grains, meat, dairy, and leafy greens.

Fat-Soluble Vitamins

Vitamin A: Crucial for vision, immune function, and skin health. Found in liver, dairy products, and orange vegetables like carrots.

Vitamin D: Important for bone health and calcium absorption. Synthesized by the skin through sunlight exposure and found in fortified foods and fatty fish.

Vitamin E: Acts as an antioxidant protecting cells from damage. Found in nuts, seeds, and vegetable oils.

Vitamin K: Essential for blood clotting and bone health. Found in leafy greens, broccoli, and Brussels sprouts.

B. Minerals

Major Minerals

Calcium: Vital for bone and teeth health, muscle function, and nerve signaling. Found in dairy products, green leafy vegetables, and fortified plant milks.

Potassium: Helps maintain fluid balance, muscle contractions, and nerve signals. Found in bananas, potatoes, and spinach.

Sodium: Important for fluid balance and nerve function, but excessive intake can lead to high blood pressure. Found in table salt and processed foods.

Magnesium: Involved in over 300 biochemical reactions, including muscle and nerve function. Found in nuts, seeds, and whole grains.

Trace Minerals

Iron: Essential for oxygen transport in the blood and energy metabolism. Found in red meat, beans, and fortified cereals.

Zinc: Important for immune function, wound healing, and DNA synthesis. Found in meat, shellfish, and legumes.

Iodine: Necessary for thyroid hormone production, which regulates metabolism. Found in iodized salt and seafood.

3. Water

A. Functions

Hydration: Maintains fluid balance and hydration essential for physiological processes.

Temperature Regulation: Helps regulate body temperature through sweating and respiration.

Nutrient Transport: Facilitates the transport of nutrients and waste products in and out of cells.

Digestion: Assists in digestion and absorption of nutrients.

B. Sources

Drinking Water: Primary source of hydration.

Foods: Fruits and vegetables with high water content, such as cucumbers, watermelon, and oranges.

4. Nutritional Requirements and Dietary Guidelines

A. Recommended Dietary Allowances (RDAs)

RDAs are set guidelines for nutrient intake levels necessary to meet the needs of most healthy individuals.

Varies by Age, Gender, and Life Stage: Nutrient needs change across different ages, genders, and during pregnancy or lactation.

B. Dietary Guidelines

Balanced Diet: Emphasizes a variety of foods from all food groups to ensure a balanced intake of nutrients.

Moderation and Portion Control: Encourages moderating intake of high-calorie, low-nutrient foods and practicing portion control.

Adequate Hydration: Promotes drinking sufficient water and incorporating hydrating foods.

5. Nutritional Deficiencies and Disorders

A. Common Nutritional Deficiencies

Iron Deficiency: Can lead to anemia, fatigue, and weakened immune function.

Vitamin D Deficiency: Can cause bone disorders such as rickets in children and osteomalacia in adults.

Vitamin A Deficiency: Can lead to vision problems, including night blindness.

B. Impact of Nutritional Imbalances

Overnutrition: Excessive intake of certain nutrients, particularly from processed foods, can lead to obesity, cardiovascular diseases, and type 2 diabetes.

Undernutrition: Inadequate intake of essential nutrients can impair growth, immune function, and overall health.

Understanding the various nutrients and their functions is fundamental to achieving optimal health and preventing nutritional deficiencies. Macronutrients (carbohydrates, proteins, and fats) provide energy and support various bodily functions, while micronutrients (vitamins and minerals) are essential for maintaining health and preventing diseases. Proper hydration is also crucial for physiological processes. By adhering to dietary guidelines and ensuring a balanced intake of nutrients, individuals can support their overall well-being and enhance their quality of life.

The Digestive System

The **digestive system** is a complex network of organs responsible for breaking down food, absorbing nutrients, and eliminating waste. It ensures that the body obtains the essential nutrients required for growth, energy, and cellular repair. This system involves both mechanical and chemical processes, allowing the body to convert large, complex food molecules into smaller, usable forms.

1. Overview of the Digestive System

The digestive system includes several organs that work in sequence to process food. The main components include the **mouth, esophagus, stomach, small intestine, large intestine, rectum,** and **anus**, as well as accessory organs like the **liver, pancreas,** and **gallbladder**.

Key Functions of the Digestive System:

Ingestion: Intake of food.

Digestion: Mechanical and chemical breakdown of food into absorbable units.

Absorption: Uptake of nutrients into the blood or lymphatic system.

Excretion: Elimination of indigestible substances as feces.

2. Major Organs of the Digestive System

A. Mouth

The digestive process begins in the **mouth**, where both mechanical and chemical digestion occur.

Mechanical Digestion:

Teeth break down food into smaller pieces through chewing (mastication), increasing the surface area for enzymes to act on.

Chemical Digestion:

Saliva, secreted by the salivary glands, contains the enzyme **amylase**, which begins the breakdown of carbohydrates (specifically starch) into simpler sugars.

The **tongue** helps mix food with saliva and forms it into a **bolus** (a soft mass) for swallowing.

B. Esophagus

After chewing, the **bolus** moves into the **esophagus**, a muscular tube that connects the mouth to the stomach. The movement of food is facilitated by a process called **peristalsis**—rhythmic contractions of the smooth muscles that propel food downward.

Esophageal Sphincter: A muscular valve that prevents the backflow of stomach contents into the esophagus, ensuring food moves in one direction.

C. Stomach

The **stomach** is a J-shaped organ that plays a crucial role in both mechanical and chemical digestion.

Mechanical Digestion:

The stomach's muscles churn and mix food, breaking it down into smaller particles and forming **chyme**, a semi-liquid mixture of partially digested food and gastric juices.

Chemical Digestion:

Gastric glands secrete **gastric juice**, which contains hydrochloric acid (HCl) and **pepsin**, an enzyme that begins the breakdown of proteins into smaller peptides.

The acidic environment (pH of 1.5-3.5) helps to denature proteins and kill bacteria.

Mucus: Protects the stomach lining from the corrosive effects of gastric acid.

Note: Little absorption occurs in the stomach, except for alcohol and certain medications.

D. Small Intestine

The **small intestine** is the primary site for nutrient digestion and absorption. It is divided into three sections: the **duodenum**, **jejunum**, and **ileum**.

Duodenum:

The first section of the small intestine where **chyme** from the stomach mixes with digestive juices from the **pancreas** and **bile** from the **liver** and **gallbladder**.

Pancreatic enzymes (like amylase, lipase, and proteases) further digest carbohydrates, fats, and proteins.

Bile emulsifies fats, breaking them into smaller droplets for easier digestion by lipase.

Jejunum and **Ileum**:

These regions are primarily responsible for the absorption of nutrients. The inner surface is lined with **villi** and **microvilli**, which increase the surface area for absorption.

Nutrients such as amino acids, simple sugars (glucose), fatty acids, vitamins, and minerals are absorbed into the bloodstream or lymphatic system through the epithelial cells of the intestinal lining.

E. Large Intestine (Colon)

The **large intestine** is responsible for absorbing water and electrolytes, forming solid waste (feces), and eliminating it from the body.

Cecum: A pouch that connects the small intestine to the large intestine.

Colon: The main part of the large intestine, divided into ascending, transverse, descending, and sigmoid regions. It absorbs water and salts, compacting the indigestible material into solid waste.

Rectum: Stores feces until they are expelled through the **anus** during defecation.

The large intestine also houses a rich microbial community, known as the **gut microbiota**, which aids in the fermentation of indigestible carbohydrates and the synthesis of certain vitamins (like vitamin K).

3. Accessory Organs

Several accessory organs play critical roles in digestion by producing and secreting essential enzymes and other substances.

A. Liver

The **liver** is the largest internal organ and performs various metabolic functions, including the production of **bile**, which aids in the digestion and absorption of fats.

Bile Production: The liver produces bile, which is stored in the gallbladder and released into the small intestine to emulsify fats.

Detoxification: The liver filters toxins and waste products from the blood, ensuring that harmful substances are removed.

B. Gallbladder

The **gallbladder** stores bile produced by the liver. During digestion, particularly after eating fatty foods, the gallbladder releases bile into the small intestine to aid in fat digestion.

C. Pancreas

The **pancreas** produces a variety of digestive enzymes that are released into the small intestine to aid in the breakdown of carbohydrates, proteins, and fats.

Pancreatic Amylase: Breaks down starches into sugars.

Lipase: Breaks down fats into fatty acids and glycerol.

Proteases (Trypsin, Chymotrypsin): Break down proteins into peptides and amino acids.

Bicarbonate: Neutralizes the acidic chyme from the stomach, creating an optimal pH for enzyme activity in the small intestine.

4. Digestive Enzymes and Their Roles

Amylase: Breaks down carbohydrates into sugars. Produced by the salivary glands and pancreas.

Pepsin: Breaks down proteins in the stomach.

Lipase: Breaks down fats into fatty acids and glycerol in the small intestine.

Trypsin and Chymotrypsin: Pancreatic enzymes that break proteins into smaller peptides.

Lactase, Maltase, and Sucrase: Enzymes in the small intestine that break down specific sugars.

5. Phases of Digestion

A. Cephalic Phase

Triggered by the sight, smell, or thought of food, the brain sends signals to prepare the stomach for food intake by stimulating gastric juice secretion.

B. Gastric Phase

Food entering the stomach triggers the release of gastric juices and begins the breakdown of proteins and fats. Stomach distension also enhances digestive activity.

C. Intestinal Phase

Chyme enters the small intestine, and hormones (like **secretin** and **cholecystokinin**) are released to regulate the flow of digestive juices and bile for continued digestion and nutrient absorption.

6. Absorption and Transport of Nutrients

Carbohydrates: Broken down into simple sugars (glucose) and absorbed into the bloodstream through the small intestine. They are transported to cells for energy or stored as glycogen in the liver and muscles.

Proteins: Digested into amino acids and absorbed in the small intestine. Amino acids are used to synthesize new proteins or converted into energy if needed.

Fats: Broken down into fatty acids and glycerol, absorbed into the lymphatic system, and transported to cells for energy or stored as adipose tissue.

7. Disorders of the Digestive System

Gastroesophageal Reflux Disease (GERD): A condition where stomach acid flows back into the esophagus, causing heartburn and damage to the esophageal lining.

Peptic Ulcers: Open sores that develop on the inner lining of the stomach or small intestine due to the erosion caused by stomach acid.

Lactose Intolerance: The inability to digest lactose (a sugar found in milk) due to a deficiency in the enzyme lactase.

Irritable Bowel Syndrome (IBS): A disorder affecting the large intestine, causing symptoms like cramping, abdominal pain, and changes in bowel habits.

Celiac Disease: An autoimmune disorder where the ingestion of gluten leads to damage in the small intestine, affecting nutrient absorption.

The digestive system is vital for converting food into the energy and nutrients required for the body's normal functioning. Each organ plays a specific role in breaking down complex food substances into absorbable molecules, ensuring that the body receives essential nutrients. Understanding the structure and function of the digestive system is crucial to maintaining good health and preventing disorders related to digestion and nutrient absorption.

Healthy Eating and Metabolism

Healthy eating and metabolism are closely interconnected aspects of human nutrition and physiology. A balanced diet supplies the necessary nutrients that our bodies need for growth, energy, and the maintenance of vital functions, while metabolism refers to the biochemical processes that convert these nutrients into energy and the building blocks required for cellular functions. Understanding the importance of healthy eating and how it relates to metabolism is essential for overall well-being.

1. Healthy Eating: Key Concepts

Healthy eating focuses on consuming a balanced diet that provides the body with essential nutrients, including carbohydrates, proteins, fats, vitamins, minerals, and water. These nutrients are vital for the maintenance of the body's physiological processes.

A. Macronutrients

Macronutrients are nutrients required in large quantities, providing the bulk of energy and building materials for the body.

Carbohydrates:

The primary source of energy for the body.

Simple carbohydrates (e.g., sugars) are quickly absorbed and provide immediate energy, while **complex carbohydrates** (e.g., starches and fiber) provide sustained energy.

Found in foods like bread, grains, fruits, and vegetables.

Carbohydrates are broken down into glucose, which fuels cellular activities, especially in the brain and muscles.

Proteins:

Essential for building and repairing tissues, proteins also play a critical role in enzyme function, hormone production, and immune system health.

Made up of amino acids, some of which are **essential** (must be obtained through the diet).

Found in meats, fish, dairy, legumes, and nuts.

Fats:

A dense energy source, fats are important for insulation, protecting organs, and helping the body absorb certain vitamins (A, D, E, and K).

Saturated fats (found in animal products) and **trans fats** (processed foods) should be consumed in moderation due to their links to heart disease.

Unsaturated fats, including monounsaturated and polyunsaturated fats (found in fish, nuts, and plant oils), are beneficial for heart health.

Fats are broken down into fatty acids and glycerol, which can be used for energy or stored for later use.

B. Micronutrients

Micronutrients include vitamins and minerals, required in smaller amounts but essential for numerous biological processes.

Vitamins:

Organic compounds that regulate metabolic functions.

Water-soluble vitamins (B-complex and C) need to be consumed regularly as they are not stored in the body, while **fat-soluble vitamins** (A, D, E, and K) are stored in body fat and the liver.

Minerals:

Inorganic elements that are vital for processes such as bone formation, muscle function, and fluid balance.

Key minerals include calcium, potassium, iron, and zinc.

C. Water

Water is essential for maintaining hydration, regulating body temperature, and enabling biochemical reactions. It plays a vital role in digestion, nutrient transport, and waste elimination. The human body is about 60% water, and sufficient intake is necessary to support all physiological processes.

D. Dietary Fiber

Dietary fiber, though indigestible, is crucial for maintaining healthy digestion. It promotes regular bowel movements, prevents constipation, and may help in regulating blood sugar levels and lowering cholesterol. Fiber is found in plant-based foods like fruits, vegetables, whole grains, and legumes.

2. Balanced Diet and Dietary Guidelines

A balanced diet emphasizes variety, moderation, and adequacy of all food groups. Different dietary guidelines are established to help individuals make healthier choices, such

as the **MyPlate** (in the U.S.) or the **Eatwell Guide** (in the U.K.), which divide food into categories and recommend portion sizes for a balanced diet. Key principles of healthy eating include:

Variety: Consuming different foods ensures intake of a wide range of nutrients.

Moderation: Balancing calorie intake with energy expenditure helps prevent overconsumption of food, particularly those high in fats and sugars.

Proportionality: Encouraging the right balance between different food groups, emphasizing fruits, vegetables, whole grains, and lean proteins while limiting fats, added sugars, and sodium.

Key components of a healthy eating pattern include:

Fruits and Vegetables: Rich in vitamins, minerals, and fiber, they reduce the risk of chronic diseases like heart disease and cancer.

Whole Grains: Provide complex carbohydrates and fiber, aiding digestion and stabilizing blood sugar levels.

Lean Proteins: Such as poultry, fish, beans, and nuts, which are essential for tissue repair and muscle growth.

Healthy Fats: Including unsaturated fats from sources like olive oil, avocados, and nuts, supporting heart health.

Limiting Processed Foods: Reducing intake of processed foods, which are often high in salt, sugar, and unhealthy fats, lowers the risk of obesity, hypertension, and type 2 diabetes.

3. Metabolism: Overview

Metabolism refers to the chemical reactions within the body's cells that convert food into energy. It is a complex process comprising **catabolism** (the breakdown of molecules to produce energy) and **anabolism** (the synthesis of compounds needed by cells). These processes ensure that the body has the energy it needs for maintenance, repair, growth, and daily functioning.

A. Basal Metabolic Rate (BMR)

Basal Metabolic Rate (BMR) is the amount of energy (calories) the body needs to maintain vital functions while at rest, such as breathing, circulation, and cell production. BMR accounts for approximately 60-75% of total daily energy expenditure.

Factors influencing BMR include:

Age: BMR decreases with age as muscle mass declines.

Gender: Males typically have a higher BMR than females due to greater muscle mass.

Body Composition: More muscle increases BMR, as muscle burns more calories at rest than fat.

Genetics: Metabolic rates can vary among individuals due to inherited traits.

Hormones: Thyroid hormones play a key role in regulating metabolism. Overproduction or underproduction can lead to hypermetabolism or hypometabolism, respectively.

B. Total Daily Energy Expenditure (TDEE)

Total Daily Energy Expenditure (TDEE) represents the total number of calories burned in a day, accounting for BMR, physical activity, and the **thermic effect of food** (the energy required to digest, absorb, and process nutrients). Physical activity can significantly increase daily caloric needs.

C. Catabolism and Anabolism

Catabolism: The breakdown of complex molecules into simpler ones, releasing energy in the form of **adenosine triphosphate (ATP)**, the energy currency of cells. For instance, during digestion, carbohydrates are broken down into glucose, fats into fatty acids, and proteins into amino acids, which are further metabolized for energy production.

Anabolism: Involves the building of complex molecules from simpler ones, using energy. This process is crucial for growth, repair, and the maintenance of tissues. For example, proteins are synthesized from amino acids to repair muscles after exercise.

4. Macronutrient Metabolism

A. Carbohydrate Metabolism

Glucose is the primary energy source for the body, especially the brain and muscles.

After ingestion, carbohydrates are broken down into glucose. This glucose can be immediately used for energy, stored as **glycogen** in the liver and muscles for short-term storage, or converted into fat for long-term storage if excess is consumed.

The **glycolysis** pathway breaks down glucose to produce ATP, particularly during high-intensity exercise.

B. Fat Metabolism

Fat provides a concentrated source of energy and can be stored in **adipose tissue** when not immediately needed.

When required for energy, fats are broken down into **fatty acids** through a process called **beta-oxidation**, which generates ATP. Fats are especially important during low-intensity, prolonged activities like walking or resting.

Excess calories from any macronutrient can be converted into fat for long-term storage.

C. Protein Metabolism

Proteins are primarily used for growth, repair, and maintenance of body tissues. However, when energy intake is insufficient, the body can use **amino acids** for energy production.

Amino acids are broken down through a process called **deamination**, where the nitrogen is removed, allowing the carbon skeleton to be used for energy or converted into glucose through **gluconeogenesis**.

5. The Role of Hormones in Metabolism

Various hormones regulate metabolic processes, ensuring energy balance and proper nutrient utilization.

Insulin: Produced by the pancreas, insulin helps regulate blood glucose levels by promoting the uptake of glucose into cells and stimulating glycogen synthesis. It also promotes fat storage and inhibits fat breakdown.

Glucagon: Also produced by the pancreas, glucagon stimulates the breakdown of glycogen into glucose during fasting or low blood sugar, providing energy to cells.

Thyroid Hormones: **Thyroxine (T4)** and **Triiodothyronine (T3)** produced by the thyroid gland increase metabolic rate by enhancing cellular oxygen consumption and energy production.

Leptin: Produced by adipose (fat) tissue, leptin helps regulate energy balance by signaling the brain to reduce appetite when fat stores are sufficient.

Ghrelin: Known as the "hunger hormone," ghrelin stimulates appetite, particularly before meals, and promotes fat storage.

6. Metabolism and Weight Management

Maintaining a balance between caloric intake and energy expenditure is crucial for healthy weight management.

Positive Energy Balance: Occurs when caloric intake exceeds expenditure, leading to weight gain as excess energy is stored as fat.

Negative Energy Balance: Occurs when caloric expenditure exceeds intake, leading to weight loss as the body uses stored fat for energy.

Weight Maintenance: Achieved when caloric intake matches expenditure, keeping body weight stable.

Factors like physical activity, diet composition, age, and metabolism play roles in managing body weight. Regular exercise, a balanced diet, and adequate sleep are essential strategies for maintaining healthy metabolism and body weight.

7. Impact of Poor Nutrition on Metabolism

Poor nutrition can adversely affect metabolism and overall health:

Malnutrition: Inadequate intake of calories, vitamins, and minerals can slow metabolism, weaken the immune system, and impair growth and development.

Obesity: Excessive caloric intake, particularly from unhealthy foods, can lead to fat accumulation, insulin resistance, and metabolic disorders such as type 2 diabetes.

Metabolic Syndrome: A cluster of conditions—including obesity, high blood pressure, high blood sugar, and abnormal cholesterol levels—linked to an increased risk of heart disease, stroke, and type 2 diabetes.

8. Strategies for Healthy Metabolism

Regular Physical Activity: Increases muscle mass and metabolic rate, particularly through resistance training and aerobic exercise.

Balanced Diet: Ensuring adequate intake of all macronutrients, vitamins, and minerals supports proper metabolic function.

Adequate Sleep: Sleep is critical for metabolic health, as insufficient sleep disrupts hormones like insulin, ghrelin, and leptin.

Hydration: Water is essential for all metabolic processes, including the breakdown of macronutrients and the transport of nutrients.

Healthy eating and metabolism are intricately linked in maintaining overall health and energy balance. A balanced diet provides the body with the necessary nutrients to fuel metabolic processes, support growth, and repair, and ensure optimal functioning of physiological systems. By understanding the importance of nutrient intake and metabolism, individuals can make informed choices that promote long-term health, manage weight, and prevent metabolic disorders.

Chapter 23: Circulatory and Respiratory Systems
Structure and Function of the Heart and Blood Vessels

Healthy eating and metabolism are closely interconnected aspects of human nutrition and physiology. A balanced diet supplies the necessary nutrients that our bodies need for growth, energy, and the maintenance of vital functions, while metabolism refers to the biochemical processes that convert these nutrients into energy and the building blocks required for cellular functions. Understanding the importance of healthy eating and how it relates to metabolism is essential for overall well-being.

1. Healthy Eating: Key Concepts

Healthy eating focuses on consuming a balanced diet that provides the body with essential nutrients, including carbohydrates, proteins, fats, vitamins, minerals, and water. These nutrients are vital for the maintenance of the body's physiological processes.

A. Macronutrients

Macronutrients are nutrients required in large quantities, providing the bulk of energy and building materials for the body.

Carbohydrates:

The primary source of energy for the body.

Simple carbohydrates (e.g., sugars) are quickly absorbed and provide immediate energy, while **complex carbohydrates** (e.g., starches and fiber) provide sustained energy.

Found in foods like bread, grains, fruits, and vegetables.

Carbohydrates are broken down into glucose, which fuels cellular activities, especially in the brain and muscles.

Proteins:

Essential for building and repairing tissues, proteins also play a critical role in enzyme function, hormone production, and immune system health.

Made up of amino acids, some of which are **essential** (must be obtained through the diet).

Found in meats, fish, dairy, legumes, and nuts.

Fats:

A dense energy source, fats are important for insulation, protecting organs, and helping the body absorb certain vitamins (A, D, E, and K).

Saturated fats (found in animal products) and **trans fats** (processed foods) should be consumed in moderation due to their links to heart disease.

Unsaturated fats, including monounsaturated and polyunsaturated fats (found in fish, nuts, and plant oils), are beneficial for heart health.

Fats are broken down into fatty acids and glycerol, which can be used for energy or stored for later use.

B. Micronutrients

Micronutrients include vitamins and minerals, required in smaller amounts but essential for numerous biological processes.

Vitamins:

Organic compounds that regulate metabolic functions.

Water-soluble vitamins (B-complex and C) need to be consumed regularly as they are not stored in the body, while **fat-soluble vitamins** (A, D, E, and K) are stored in body fat and the liver.

Minerals:

Inorganic elements that are vital for processes such as bone formation, muscle function, and fluid balance.

Key minerals include calcium, potassium, iron, and zinc.

C. Water

Water is essential for maintaining hydration, regulating body temperature, and enabling biochemical reactions. It plays a vital role in digestion, nutrient transport, and waste elimination. The human body is about 60% water, and sufficient intake is necessary to support all physiological processes.

D. Dietary Fiber

Dietary fiber, though indigestible, is crucial for maintaining healthy digestion. It promotes regular bowel movements, prevents constipation, and may help in regulating blood sugar levels and lowering cholesterol. Fiber is found in plant-based foods like fruits, vegetables, whole grains, and legumes.

2. Balanced Diet and Dietary Guidelines

A balanced diet emphasizes variety, moderation, and adequacy of all food groups. Different dietary guidelines are established to help individuals make healthier choices, such

as the **MyPlate** (in the U.S.) or the **Eatwell Guide** (in the U.K.), which divide food into categories and recommend portion sizes for a balanced diet. Key principles of healthy eating include:

Variety: Consuming different foods ensures intake of a wide range of nutrients.

Moderation: Balancing calorie intake with energy expenditure helps prevent overconsumption of food, particularly those high in fats and sugars.

Proportionality: Encouraging the right balance between different food groups, emphasizing fruits, vegetables, whole grains, and lean proteins while limiting fats, added sugars, and sodium.

Key components of a healthy eating pattern include:

Fruits and Vegetables: Rich in vitamins, minerals, and fiber, they reduce the risk of chronic diseases like heart disease and cancer.

Whole Grains: Provide complex carbohydrates and fiber, aiding digestion and stabilizing blood sugar levels.

Lean Proteins: Such as poultry, fish, beans, and nuts, which are essential for tissue repair and muscle growth.

Healthy Fats: Including unsaturated fats from sources like olive oil, avocados, and nuts, supporting heart health.

Limiting Processed Foods: Reducing intake of processed foods, which are often high in salt, sugar, and unhealthy fats, lowers the risk of obesity, hypertension, and type 2 diabetes.

3. Metabolism: Overview

Metabolism refers to the chemical reactions within the body's cells that convert food into energy. It is a complex process comprising **catabolism** (the breakdown of molecules to produce energy) and **anabolism** (the synthesis of compounds needed by cells). These processes ensure that the body has the energy it needs for maintenance, repair, growth, and daily functioning.

A. Basal Metabolic Rate (BMR)

Basal Metabolic Rate (BMR) is the amount of energy (calories) the body needs to maintain vital functions while at rest, such as breathing, circulation, and cell production. BMR accounts for approximately 60-75% of total daily energy expenditure.

Factors influencing BMR include:

Age: BMR decreases with age as muscle mass declines.

Gender: Males typically have a higher BMR than females due to greater muscle mass.

Body Composition: More muscle increases BMR, as muscle burns more calories at rest than fat.

Genetics: Metabolic rates can vary among individuals due to inherited traits.

Hormones: Thyroid hormones play a key role in regulating metabolism. Overproduction or underproduction can lead to hypermetabolism or hypometabolism, respectively.

B. Total Daily Energy Expenditure (TDEE)

Total Daily Energy Expenditure (TDEE) represents the total number of calories burned in a day, accounting for BMR, physical activity, and the **thermic effect of food** (the energy required to digest, absorb, and process nutrients). Physical activity can significantly increase daily caloric needs.

C. Catabolism and Anabolism

Catabolism: The breakdown of complex molecules into simpler ones, releasing energy in the form of **adenosine triphosphate (ATP)**, the energy currency of cells. For instance, during digestion, carbohydrates are broken down into glucose, fats into fatty acids, and proteins into amino acids, which are further metabolized for energy production.

Anabolism: Involves the building of complex molecules from simpler ones, using energy. This process is crucial for growth, repair, and the maintenance of tissues. For example, proteins are synthesized from amino acids to repair muscles after exercise.

4. Macronutrient Metabolism

A. Carbohydrate Metabolism

Glucose is the primary energy source for the body, especially the brain and muscles.

After ingestion, carbohydrates are broken down into glucose. This glucose can be immediately used for energy, stored as **glycogen** in the liver and muscles for short-term storage, or converted into fat for long-term storage if excess is consumed.

The **glycolysis** pathway breaks down glucose to produce ATP, particularly during high-intensity exercise.

B. Fat Metabolism

Fat provides a concentrated source of energy and can be stored in **adipose tissue** when not immediately needed.

When required for energy, fats are broken down into **fatty acids** through a process called **beta-oxidation**, which generates ATP. Fats are especially important during low-intensity, prolonged activities like walking or resting.

Excess calories from any macronutrient can be converted into fat for long-term storage.

C. Protein Metabolism

Proteins are primarily used for growth, repair, and maintenance of body tissues. However, when energy intake is insufficient, the body can use **amino acids** for energy production.

Amino acids are broken down through a process called **deamination**, where the nitrogen is removed, allowing the carbon skeleton to be used for energy or converted into glucose through **gluconeogenesis**.

5. The Role of Hormones in Metabolism

Various hormones regulate metabolic processes, ensuring energy balance and proper nutrient utilization.

Insulin: Produced by the pancreas, insulin helps regulate blood glucose levels by promoting the uptake of glucose into cells and stimulating glycogen synthesis. It also promotes fat storage and inhibits fat breakdown.

Glucagon: Also produced by the pancreas, glucagon stimulates the breakdown of glycogen into glucose during fasting or low blood sugar, providing energy to cells.

Thyroid Hormones: **Thyroxine (T4)** and **Triiodothyronine (T3)** produced by the thyroid gland increase metabolic rate by enhancing cellular oxygen consumption and energy production.

Leptin: Produced by adipose (fat) tissue, leptin helps regulate energy balance by signaling the brain to reduce appetite when fat stores are sufficient.

Ghrelin: Known as the "hunger hormone," ghrelin stimulates appetite, particularly before meals, and promotes fat storage.

6. Metabolism and Weight Management

Maintaining a balance between caloric intake and energy expenditure is crucial for healthy weight management.

Positive Energy Balance: Occurs when caloric intake exceeds expenditure, leading to weight gain as excess energy is stored as fat.

Negative Energy Balance: Occurs when caloric expenditure exceeds intake, leading to weight loss as the body uses stored fat for energy.

Weight Maintenance: Achieved when caloric intake matches expenditure, keeping body weight stable.

Factors like physical activity, diet composition, age, and metabolism play roles in managing body weight. Regular exercise, a balanced diet, and adequate sleep are essential strategies for maintaining healthy metabolism and body weight.

7. Impact of Poor Nutrition on Metabolism

Poor nutrition can adversely affect metabolism and overall health:

Malnutrition: Inadequate intake of calories, vitamins, and minerals can slow metabolism, weaken the immune system, and impair growth and development.

Obesity: Excessive caloric intake, particularly from unhealthy foods, can lead to fat accumulation, insulin resistance, and metabolic disorders such as type 2 diabetes.

Metabolic Syndrome: A cluster of conditions—including obesity, high blood pressure, high blood sugar, and abnormal cholesterol levels—linked to an increased risk of heart disease, stroke, and type 2 diabetes.

8. Strategies for Healthy Metabolism

Regular Physical Activity: Increases muscle mass and metabolic rate, particularly through resistance training and aerobic exercise.

Balanced Diet: Ensuring adequate intake of all macronutrients, vitamins, and minerals supports proper metabolic function.

Adequate Sleep: Sleep is critical for metabolic health, as insufficient sleep disrupts hormones like insulin, ghrelin, and leptin.

Hydration: Water is essential for all metabolic processes, including the breakdown of macronutrients and the transport of nutrients.

Healthy eating and metabolism are intricately linked in maintaining overall health and energy balance. A balanced diet provides the body with the necessary nutrients to fuel metabolic processes, support growth, and repair, and ensure optimal functioning of physiological systems. By understanding the importance of nutrient intake and metabolism, individuals can make informed choices that promote long-term health, manage weight, and prevent metabolic disorders.

Blood and Lymphatic System

The circulatory system is responsible for transporting nutrients, gases, hormones, and waste products throughout the body. It consists of two interconnected systems: the **blood circulatory system** and the **lymphatic system**. Together, they maintain homeostasis by regulating fluid balance, immunity, and nutrient distribution.

1. Blood

Blood is a specialized connective tissue that plays a vital role in transporting essential substances throughout the body. It comprises **plasma** (the liquid portion) and **formed elements** (cells and cell fragments). Blood has various functions, including:

Transport of gases, nutrients, and waste products: Blood carries oxygen from the lungs to tissues and carbon dioxide from tissues to the lungs. It also transports nutrients from the digestive system and waste products to excretory organs.

Regulation of body temperature: Blood helps maintain a stable internal temperature by distributing heat generated in metabolic processes.

Immunity and protection: Blood contains immune cells and antibodies that defend against pathogens.

Clotting: Blood can coagulate to prevent excessive bleeding after injury.

A. Composition of Blood

Plasma:

Plasma makes up about 55% of blood and is primarily water (about 90%), but it also contains dissolved proteins, electrolytes, hormones, nutrients, and waste products.

Plasma proteins like **albumin** (maintains osmotic pressure), **fibrinogen** (involved in clotting), and **globulins** (antibodies for immune response) play critical roles in various physiological processes.

Formed Elements: The solid components of blood make up about 45% and include:

Erythrocytes (Red Blood Cells): Erythrocytes are biconcave cells responsible for transporting oxygen and carbon dioxide. They contain **hemoglobin**, a protein that binds oxygen. Red blood cells lack a nucleus and mitochondria, allowing more room for hemoglobin, and have a lifespan of about 120 days.

Leukocytes (White Blood Cells): These cells are crucial for the immune response, defending the body against infections and foreign invaders. There are five types of leukocytes:

Neutrophils: The most abundant, they are involved in engulfing bacteria through phagocytosis.

Lymphocytes: Include **B cells** (which produce antibodies) and **T cells** (which attack infected or cancerous cells).

Monocytes: Become **macrophages** that engulf pathogens and dead cells.

Eosinophils: Combat parasites and are involved in allergic responses.

Basophils: Release histamine during allergic reactions and inflammation.

Platelets (Thrombocytes): Platelets are cell fragments involved in blood clotting. When a blood vessel is injured, platelets aggregate to form a temporary plug, and along with clotting factors, they help form a stable clot to prevent excessive blood loss.

B. Blood Clotting (Hemostasis)

Blood clotting, or hemostasis, is the process that prevents excessive blood loss after an injury. It involves three major steps:

Vascular Spasm: The immediate constriction of blood vessels reduces blood flow.

Platelet Plug Formation: Platelets adhere to the site of injury, releasing chemicals that attract more platelets, forming a temporary plug.

Coagulation: A cascade of clotting factors leads to the conversion of **fibrinogen** (a plasma protein) into **fibrin**, which forms a mesh that stabilizes the platelet plug and creates a clot.

2. The Lymphatic System

The lymphatic system is closely connected to the circulatory system and plays a crucial role in fluid balance, fat absorption, and immunity. It consists of **lymph, lymphatic vessels, lymph nodes**, and **lymphoid organs** (such as the spleen and thymus).

A. Functions of the Lymphatic System

Fluid Balance: The lymphatic system helps maintain the body's fluid levels by collecting excess interstitial fluid (fluid that leaks out of capillaries into tissues) and returning it to the bloodstream. If this fluid were not removed, it would accumulate, leading to swelling, or **edema**.

Fat Absorption: In the small intestine, specialized lymphatic capillaries called **lacteals** absorb dietary fats and fat-soluble vitamins (A, D, E, K) from digested food and transport them to the bloodstream. The lymph enriched with fats is called **chyle**.

Immune Response: The lymphatic system plays a key role in immunity by filtering pathogens and foreign particles through the **lymph nodes**. Lymph nodes contain large numbers of lymphocytes and macrophages that trap and destroy harmful substances, including bacteria, viruses, and cancer cells.

B. Components of the Lymphatic System

Lymph:

Lymph is a clear fluid derived from the interstitial fluid that surrounds cells in tissues. It contains water, proteins, waste products, immune cells, and fats. Once the interstitial fluid enters the lymphatic vessels, it is called lymph.

Lymph flows through the lymphatic vessels and is eventually returned to the blood via the **subclavian veins**.

Lymphatic Vessels:

Lymphatic vessels form a one-way system that begins with **lymphatic capillaries** in the tissues, which absorb interstitial fluid. These capillaries merge into larger vessels, similar to veins, that transport lymph.

Lymphatic vessels contain valves to prevent backflow and ensure that lymph flows toward the heart. The movement of lymph is aided by the contraction of surrounding skeletal muscles and changes in pressure within the thorax during breathing.

Lymph Nodes:

Lymph nodes are small, bean-shaped structures located along lymphatic vessels. They act as filters for lymph, trapping bacteria, viruses, and other foreign particles, which are then destroyed by immune cells.

Common areas where lymph nodes are found include the neck (cervical lymph nodes), armpits (axillary lymph nodes), and groin (inguinal lymph nodes).

Lymphoid Organs:

Spleen: The largest lymphatic organ, the spleen filters blood, removes old or damaged red blood cells, and helps fight infection by producing and storing lymphocytes.

Thymus: Located near the heart, the thymus is where **T cells** (a type of lymphocyte) mature. The thymus is particularly active during childhood but shrinks as a person ages.

Tonsils: Located at the back of the throat, tonsils help trap pathogens entering the body through the mouth and nose.

C. The Immune System and Lymphatic Function

The lymphatic system works closely with the immune system to defend the body against infections and diseases. Lymph nodes and other lymphoid organs act as sites for immune surveillance, where **lymphocytes** (B and T cells) monitor the lymph for foreign invaders and initiate immune responses when necessary.

B cells: Produce antibodies that bind to pathogens and mark them for destruction.

T cells: Directly attack infected or cancerous cells, and help regulate immune responses.

The lymphatic system also serves as a route for the spread of cancer, as cancer cells can enter lymph vessels and travel to lymph nodes. This is why lymph nodes are often checked during cancer diagnoses.

D. Lymphatic Disorders

Several disorders can affect the lymphatic system, including:

Lymphedema: A condition where lymphatic fluid accumulates in tissues, leading to swelling, often in the arms or legs. It can be caused by damage to lymphatic vessels or nodes, infection, or surgery (e.g., lymph node removal during cancer treatment).

Lymphoma: A type of cancer that begins in lymphocytes (white blood cells) and affects lymph nodes and other lymphoid tissues. Hodgkin's and non-Hodgkin's lymphoma are two main types.

Infections: Pathogens such as bacteria, viruses, and parasites can infect the lymphatic system, leading to conditions like **lymphangitis** (inflammation of lymph vessels) or **lymphadenitis** (inflammation of lymph nodes).

3. Interaction Between Blood and Lymphatic Systems

While the blood and lymphatic systems operate independently, they are interconnected in maintaining fluid balance and immunity. The excess fluid from blood capillaries that enters tissues becomes interstitial fluid, which is then collected by lymphatic vessels, filtered by lymph nodes, and eventually returned to the blood. Moreover, the lymphatic system assists in the immune response by transporting immune cells and filtering pathogens out of the lymph.

The blood circulatory and lymphatic systems are essential for maintaining homeostasis, immunity, and proper physiological functioning. Blood circulates nutrients, gases, and waste products throughout the body, while the lymphatic system ensures fluid balance, absorbs fats, and plays a critical role in immune defense. Together, these systems are vital for health and survival, working in tandem to protect and nourish the body.

The Respiratory System and Gas Exchange

The circulatory system is responsible for transporting nutrients, gases, hormones, and waste products throughout the body. It consists of two interconnected systems: the **blood circulatory system** and the **lymphatic system**. Together, they maintain homeostasis by regulating fluid balance, immunity, and nutrient distribution.

1. Blood

Blood is a specialized connective tissue that plays a vital role in transporting essential substances throughout the body. It comprises **plasma** (the liquid portion) and **formed elements** (cells and cell fragments). Blood has various functions, including:

Transport of gases, nutrients, and waste products: Blood carries oxygen from the lungs to tissues and carbon dioxide from tissues to the lungs. It also transports nutrients from the digestive system and waste products to excretory organs.

Regulation of body temperature: Blood helps maintain a stable internal temperature by distributing heat generated in metabolic processes.

Immunity and protection: Blood contains immune cells and antibodies that defend against pathogens.

Clotting: Blood can coagulate to prevent excessive bleeding after injury.

A. Composition of Blood

Plasma:

Plasma makes up about 55% of blood and is primarily water (about 90%), but it also contains dissolved proteins, electrolytes, hormones, nutrients, and waste products.

Plasma proteins like **albumin** (maintains osmotic pressure), **fibrinogen** (involved in clotting), and **globulins** (antibodies for immune response) play critical roles in various physiological processes.

Formed Elements: The solid components of blood make up about 45% and include:

Erythrocytes (Red Blood Cells): Erythrocytes are biconcave cells responsible for transporting oxygen and carbon dioxide. They contain **hemoglobin**, a protein that binds oxygen. Red blood cells lack a nucleus and mitochondria, allowing more room for hemoglobin, and have a lifespan of about 120 days.

Leukocytes (White Blood Cells): These cells are crucial for the immune response, defending the body against infections and foreign invaders. There are five types of leukocytes:

Neutrophils: The most abundant, they are involved in engulfing bacteria through phagocytosis.

Lymphocytes: Include **B cells** (which produce antibodies) and **T cells** (which attack infected or cancerous cells).

Monocytes: Become **macrophages** that engulf pathogens and dead cells.

Eosinophils: Combat parasites and are involved in allergic responses.

Basophils: Release histamine during allergic reactions and inflammation.

Platelets (Thrombocytes): Platelets are cell fragments involved in blood clotting. When a blood vessel is injured, platelets aggregate to form a temporary plug, and along with clotting factors, they help form a stable clot to prevent excessive blood loss.

B. Blood Clotting (Hemostasis)

Blood clotting, or hemostasis, is the process that prevents excessive blood loss after an injury. It involves three major steps:

Vascular Spasm: The immediate constriction of blood vessels reduces blood flow.

Platelet Plug Formation: Platelets adhere to the site of injury, releasing chemicals that attract more platelets, forming a temporary plug.

Coagulation: A cascade of clotting factors leads to the conversion of **fibrinogen** (a plasma protein) into **fibrin**, which forms a mesh that stabilizes the platelet plug and creates a clot.

2. The Lymphatic System

The lymphatic system is closely connected to the circulatory system and plays a crucial role in fluid balance, fat absorption, and immunity. It consists of **lymph, lymphatic vessels, lymph nodes**, and **lymphoid organs** (such as the spleen and thymus).

A. Functions of the Lymphatic System

Fluid Balance: The lymphatic system helps maintain the body's fluid levels by collecting excess interstitial fluid (fluid that leaks out of capillaries into tissues) and returning it to the bloodstream. If this fluid were not removed, it would accumulate, leading to swelling, or **edema**.

Fat Absorption: In the small intestine, specialized lymphatic capillaries called **lacteals** absorb dietary fats and fat-soluble vitamins (A, D, E, K) from digested food and transport them to the bloodstream. The lymph enriched with fats is called **chyle**.

Immune Response: The lymphatic system plays a key role in immunity by filtering pathogens and foreign particles through the **lymph nodes**. Lymph nodes contain large numbers of lymphocytes and macrophages that trap and destroy harmful substances, including bacteria, viruses, and cancer cells.

B. Components of the Lymphatic System

Lymph:

Lymph is a clear fluid derived from the interstitial fluid that surrounds cells in tissues. It contains water, proteins, waste products, immune cells, and fats. Once the interstitial fluid enters the lymphatic vessels, it is called lymph.

Lymph flows through the lymphatic vessels and is eventually returned to the blood via the **subclavian veins**.

Lymphatic Vessels:

Lymphatic vessels form a one-way system that begins with **lymphatic capillaries** in the tissues, which absorb interstitial fluid. These capillaries merge into larger vessels, similar to veins, that transport lymph.

Lymphatic vessels contain valves to prevent backflow and ensure that lymph flows toward the heart. The movement of lymph is aided by the contraction of surrounding skeletal muscles and changes in pressure within the thorax during breathing.

Lymph Nodes:

Lymph nodes are small, bean-shaped structures located along lymphatic vessels. They act as filters for lymph, trapping bacteria, viruses, and other foreign particles, which are then destroyed by immune cells.

Common areas where lymph nodes are found include the neck (cervical lymph nodes), armpits (axillary lymph nodes), and groin (inguinal lymph nodes).

Lymphoid Organs:

Spleen: The largest lymphatic organ, the spleen filters blood, removes old or damaged red blood cells, and helps fight infection by producing and storing lymphocytes.

Thymus: Located near the heart, the thymus is where **T cells** (a type of lymphocyte) mature. The thymus is particularly active during childhood but shrinks as a person ages.

Tonsils: Located at the back of the throat, tonsils help trap pathogens entering the body through the mouth and nose.

C. The Immune System and Lymphatic Function

The lymphatic system works closely with the immune system to defend the body against infections and diseases. Lymph nodes and other lymphoid organs act as sites for immune surveillance, where **lymphocytes** (B and T cells) monitor the lymph for foreign invaders and initiate immune responses when necessary.

B cells: Produce antibodies that bind to pathogens and mark them for destruction.

T cells: Directly attack infected or cancerous cells, and help regulate immune responses.

The lymphatic system also serves as a route for the spread of cancer, as cancer cells can enter lymph vessels and travel to lymph nodes. This is why lymph nodes are often checked during cancer diagnoses.

D. Lymphatic Disorders

Several disorders can affect the lymphatic system, including:

Lymphedema: A condition where lymphatic fluid accumulates in tissues, leading to swelling, often in the arms or legs. It can be caused by damage to lymphatic vessels or nodes, infection, or surgery (e.g., lymph node removal during cancer treatment).

Lymphoma: A type of cancer that begins in lymphocytes (white blood cells) and affects lymph nodes and other lymphoid tissues. Hodgkin's and non-Hodgkin's lymphoma are two main types.

Infections: Pathogens such as bacteria, viruses, and parasites can infect the lymphatic system, leading to conditions like **lymphangitis** (inflammation of lymph vessels) or **lymphadenitis** (inflammation of lymph nodes).

3. Interaction Between Blood and Lymphatic Systems

While the blood and lymphatic systems operate independently, they are interconnected in maintaining fluid balance and immunity. The excess fluid from blood capillaries that enters tissues becomes interstitial fluid, which is then collected by lymphatic vessels, filtered by lymph nodes, and eventually returned to the blood. Moreover, the lymphatic system assists in the immune response by transporting immune cells and filtering pathogens out of the lymph.

The blood circulatory and lymphatic systems are essential for maintaining homeostasis, immunity, and proper physiological functioning. Blood circulates nutrients, gases, and waste products throughout the body, while the lymphatic system ensures fluid balance, absorbs fats, and plays a critical role in immune defense. Together, these systems are vital for health and survival, working in tandem to protect and nourish the body.

Chapter 24: The Nervous and Endocrine Systems

Neurons and Nerve Impulses

The nervous system plays a fundamental role in controlling and regulating the body's functions by transmitting electrical and chemical signals between different parts of the body. The basic unit of the nervous system is the **neuron**, which is responsible for receiving and transmitting these signals in the form of **nerve impulses**. Neurons and nerve impulses work together to enable complex processes such as movement, sensation, thought, and emotion.

1. Structure of Neurons

Neurons are specialized cells designed to transmit electrical signals efficiently. They are composed of three main parts:

A. Cell Body (Soma)

The **cell body** is the central part of the neuron that contains the **nucleus** and most of the cell's organelles, including mitochondria, ribosomes, and the rough endoplasmic reticulum. The cell body is responsible for maintaining the neuron's health and functionality by producing proteins and energy. The nucleus contains the genetic material (DNA) that controls the cell's activities.

B. Dendrites

Dendrites are short, branching structures that extend from the cell body. They receive signals (usually chemical signals called neurotransmitters) from other neurons or sensory receptors and convert them into electrical signals. These signals are then transmitted to the cell body. The large surface area of dendrites allows neurons to receive input from many other cells, which is crucial for the complex processing of information.

C. Axon

The **axon** is a long, tube-like extension of the cell that carries electrical signals (nerve impulses) away from the cell body toward other neurons, muscles, or glands. The axon can be very short or extremely long, depending on the type of neuron. Axons are often insulated by a **myelin sheath**, which is made of **Schwann cells** (in the peripheral nervous system) or **oligodendrocytes** (in the central nervous system). The myelin sheath allows for faster transmission of impulses by preventing the loss of electrical signal.

Nodes of Ranvier: These are gaps in the myelin sheath along the axon where the action potential is regenerated. The signal "jumps" from one node to the next, a process known as **saltatory conduction**.

D. Synaptic Terminals

At the end of the axon are **synaptic terminals** (also known as axon terminals or boutons). These terminals form connections, or **synapses**, with other neurons, muscles, or glands. When the nerve impulse reaches the synaptic terminal, it triggers the release of chemical messengers called **neurotransmitters** into the synaptic cleft (the space between neurons). Neurotransmitters then bind to receptors on the next neuron or target cell, allowing the signal to continue.

2. Types of Neurons

Neurons can be categorized into three main types based on their function:

Sensory Neurons (Afferent Neurons): These neurons transmit signals from sensory receptors (e.g., in the skin, eyes, or ears) to the central nervous system (CNS). They detect changes in the environment, such as light, sound, or temperature.

Motor Neurons (Efferent Neurons): These neurons carry signals from the CNS to muscles or glands, initiating a response such as muscle contraction or hormone secretion.

Interneurons: These neurons are found primarily in the brain and spinal cord and connect sensory neurons to motor neurons. They process and interpret information received from sensory neurons and decide on an appropriate response.

3. The Resting Membrane Potential

Neurons maintain a **resting membrane potential** when they are not actively transmitting signals. The resting membrane potential is the electrical difference between the inside and outside of the neuron, with the inside being more negative relative to the outside, usually around **-70 millivolts (mV)**. This is due to the distribution of ions across the neuron's membrane.

Sodium-Potassium Pump: The resting membrane potential is maintained by the **sodium-potassium pump**, a membrane protein that actively transports **three sodium ions (Na$^+$)** out of the cell for every **two potassium ions (K$^+$)** it pumps into the cell. This creates a higher concentration of sodium outside the cell and a higher concentration of potassium inside.

Ion Channels: The membrane is more permeable to potassium than to sodium due to **potassium leak channels**. As a result, more potassium diffuses out of the cell than sodium diffuses in, which helps maintain the negative charge inside the neuron.

4. Action Potentials: How Nerve Impulses Are Generated

A **nerve impulse**, or **action potential**, is a rapid, temporary change in the electrical charge of the neuron's membrane. This is how neurons communicate and transmit signals over long distances.

A. Depolarization

When a neuron is stimulated by a signal (e.g., a neurotransmitter binding to its receptor), sodium channels open, allowing sodium ions to flow into the neuron. This causes the inside of the neuron to become more positive, a process called **depolarization**. If the membrane potential reaches a threshold (about **-55 mV**), an action potential is triggered.

B. Action Potential

Once the threshold is reached, more sodium channels open, and the membrane potential rapidly rises to around **+30 mV**. This spike in electrical activity constitutes the action potential. The action potential is an **all-or-nothing** event: if the threshold is reached, the action potential occurs; if not, it does not.

C. Repolarization

After the peak of the action potential, sodium channels close, and **potassium channels** open, allowing potassium ions to flow out of the cell. This restores the negative charge inside the neuron, a process known as **repolarization**.

D. Hyperpolarization and Refractory Period

Sometimes, the outflow of potassium causes the membrane potential to become more negative than the resting potential, a phase called **hyperpolarization**. During this **refractory period**, the neuron cannot fire another action potential. The sodium-potassium pump restores the resting membrane potential by moving sodium ions out and potassium ions back into the cell.

5. Propagation of the Action Potential

Once generated, the action potential must travel along the length of the axon to the synaptic terminals. This process is known as **propagation**.

A. Continuous Conduction

In unmyelinated axons, the action potential travels down the entire length of the axon membrane in a continuous wave. This is slower because the signal must be regenerated at every point along the axon.

B. Saltatory Conduction

In myelinated axons, the action potential jumps from one **node of Ranvier** to the next. Myelin acts as an insulator, preventing ion flow across the membrane except at the nodes. This allows the signal to move much more quickly, as it doesn't need to be regenerated continuously.

6. Synaptic Transmission

The action potential reaches the end of the axon at the **synaptic terminals**, where it triggers the release of neurotransmitters. This is how the signal is passed from one neuron to the next or to a target cell (e.g., a muscle or gland).

A. Electrical Synapses

At **electrical synapses**, ions flow directly between cells through **gap junctions**, allowing the action potential to pass quickly from one neuron to another. Electrical synapses are rare in the human nervous system but are found in the heart and some smooth muscle.

B. Chemical Synapses

Most synapses in the human nervous system are **chemical synapses**. In this case, the action potential causes **calcium ions** to enter the synaptic terminal. This influx of calcium triggers the release of neurotransmitters stored in **synaptic vesicles**. The neurotransmitters cross the **synaptic cleft** (the small gap between neurons) and bind to receptors on the postsynaptic neuron or target cell.

Excitatory Neurotransmitters: These increase the likelihood that the postsynaptic neuron will fire an action potential by depolarizing its membrane.

Inhibitory Neurotransmitters: These decrease the likelihood of an action potential by hyperpolarizing the postsynaptic membrane.

7. Types of Neurotransmitters

Different neurotransmitters have specific roles in the nervous system:

Acetylcholine: Involved in muscle contraction and found at neuromuscular junctions.

Dopamine: Associated with reward, motivation, and motor control.

Serotonin: Regulates mood, sleep, and appetite.

GABA (Gamma-Aminobutyric Acid): The primary inhibitory neurotransmitter in the brain.

8. Disorders of Neurons and Nerve Impulses

Several neurological disorders result from abnormalities in neurons or their ability to transmit nerve impulses:

Multiple Sclerosis (MS): An autoimmune disease where the immune system attacks the myelin sheath, disrupting saltatory conduction and causing a range of neurological symptoms.

Epilepsy: A disorder characterized by abnormal electrical activity in the brain, leading to seizures.

Parkinson's Disease: A degenerative disorder where neurons that produce dopamine die, affecting motor control.

Neurons and nerve impulses form the foundation of the nervous system. The process of transmitting electrical signals along neurons, communicating at synapses through neurotransmitters, and propagating these signals throughout the body is essential for controlling and coordinating bodily functions. Understanding this complex system provides insight into how the brain processes information, regulates behavior, and maintains homeostasis.

The Brain and Spinal Cord

The central nervous system (CNS) consists of the **brain** and **spinal cord**. It is responsible for integrating sensory information, coordinating responses, and regulating various bodily functions. The CNS acts as the command center for the body, processing information from the environment and generating appropriate responses. Understanding the structure and function of the brain and spinal cord is crucial for comprehending how the nervous system controls and coordinates bodily functions.

1. Structure and Function of the Brain

The brain is the most complex organ in the human body and is divided into several regions, each with specialized functions. It is protected by the skull and covered by three protective membranes called **meninges**. The brain is also cushioned by cerebrospinal fluid (CSF), which provides a buoyant environment and helps to protect the brain from injury.

A. Major Regions of the Brain

Cerebrum

Structure: The cerebrum is the largest part of the brain, divided into two hemispheres (left and right) and further subdivided into **lobes**.

Function: It is responsible for higher brain functions such as sensory perception, voluntary motor activities, reasoning, problem-solving, and emotions.

Lobes:

Frontal Lobe: Involved in decision-making, planning, problem-solving, and controlling voluntary movements. It contains the **primary motor cortex**, which controls movement.

Parietal Lobe: Processes sensory information related to touch, temperature, and pain. It houses the **primary somatosensory cortex**, which interprets sensory data from the body.

Temporal Lobe: Associated with processing auditory information and is involved in memory and language comprehension. The **hippocampus**, crucial for memory formation, is located here.

Occipital Lobe: Primarily responsible for processing visual information. It contains the **primary visual cortex**, which interprets visual signals from the eyes.

Diencephalon

Structure: Located deep within the brain, it includes structures such as the **thalamus** and **hypothalamus**.

Function:

Thalamus: Acts as a relay station for sensory information, directing it to the appropriate areas of the cerebrum.

Hypothalamus: Regulates homeostatic functions, such as temperature control, hunger, thirst, and circadian rhythms. It also links the nervous system to the endocrine system via the **pituitary gland**.

Brainstem

Structure: Consists of the **midbrain, pons,** and **medulla oblongata**.

Function: Controls basic life functions such as breathing, heart rate, and blood pressure. It also acts as a conduit for nerve signals between the brain and the spinal cord.

Midbrain: Involved in visual and auditory reflexes and controls some motor functions.

Pons: Connects the cerebrum to the cerebellum and is involved in regulating sleep and respiration.

Medulla Oblongata: Controls autonomic functions such as heart rate, breathing, and blood pressure.

Cerebellum

Structure: Located under the cerebrum at the back of the brain.

Function: Coordinates voluntary movements, maintains balance, and ensures smooth and accurate motor control. It integrates sensory information to adjust motor activities.

B. Protective Structures

Meninges: Three layers of protective tissue covering the brain and spinal cord:

Dura Mater: The outermost, thick, and durable layer.

Arachnoid Mater: The middle layer, which is web-like and contains cerebrospinal fluid.

Pia Mater: The innermost layer, which is thin and adheres closely to the brain's surface.

Cerebrospinal Fluid (CSF): Produced in the **choroid plexuses** of the ventricles, CSF circulates through the ventricles and around the brain and spinal cord. It provides cushioning, removes waste, and maintains a stable environment for the CNS.

2. Structure and Function of the Spinal Cord

The spinal cord extends from the base of the brain (at the **foramen magnum**) down the vertebral column. It acts as a major conduit for signals traveling between the brain and the rest of the body and coordinates reflexes.

A. Structure of the Spinal Cord

Gray Matter: The inner, butterfly-shaped region of the spinal cord composed primarily of neuron cell bodies and dendrites. It is divided into **dorsal (posterior) horns** (sensory processing) and **ventral (anterior) horns** (motor control).

White Matter: Surrounds the gray matter and consists of myelinated axons organized into **ascending** (sensory) and **descending** (motor) tracts. These tracts transmit signals to and from the brain.

Spinal Nerves: Emerge from the spinal cord and are organized into 31 pairs, each corresponding to a segment of the spinal cord. They are mixed nerves, containing both sensory and motor fibers.

B. Functions of the Spinal Cord

Conduit for Nerve Signals: Transmits sensory information from the body to the brain and motor commands from the brain to the body. Ascending tracts carry sensory information such as touch and temperature to the brain, while descending tracts carry motor commands from the brain to muscles and glands.

Reflexes: The spinal cord is involved in reflex actions, which are rapid, automatic responses to stimuli. Reflexes occur without conscious brain involvement and are processed at the spinal cord level. For example, the **withdrawal reflex** (e.g., pulling your hand away from a hot surface) is an involuntary response to protect the body from harm.

3. Integration of Brain and Spinal Cord

The brain and spinal cord work together to process information and coordinate responses:

Ascending Pathways: Sensory information from the body is relayed through the spinal cord to the brain, where it is interpreted and integrated.

Descending Pathways: Motor commands generated by the brain are transmitted down the spinal cord to the appropriate muscles and glands.

4. Common Disorders and Injuries

Disorders and injuries of the brain and spinal cord can have significant impacts on function and health:

Stroke: A sudden loss of blood flow to a part of the brain, leading to damage of brain tissue and loss of function. Symptoms depend on the affected brain region and may include paralysis, speech difficulties, and loss of cognitive abilities.

Spinal Cord Injury: Damage to the spinal cord can result from trauma (e.g., car accidents) or diseases (e.g., multiple sclerosis). The effects of spinal cord injury vary based on the location and severity of the damage and may include loss of sensation, paralysis, or autonomic dysfunction.

Alzheimer's Disease: A neurodegenerative disorder characterized by progressive memory loss and cognitive decline. It involves the accumulation of abnormal protein aggregates in the brain, leading to neuronal death and dysfunction.

Parkinson's Disease: A movement disorder caused by the degeneration of dopamine-producing neurons in the basal ganglia. It leads to tremors, stiffness, and difficulty with movement and coordination.

Multiple Sclerosis (MS): An autoimmune disease where the immune system attacks the myelin sheath of neurons in the central nervous system, leading to a range of neurological symptoms including muscle weakness, vision problems, and impaired coordination.

The brain and spinal cord form the central nervous system, which is essential for processing information and coordinating bodily functions. Understanding their structure and functions is crucial for diagnosing and treating neurological disorders, as well as for appreciating the complexity of how the nervous system controls and regulates the body's activities.

Hormones and Glands

The endocrine system works in conjunction with the nervous system to regulate and coordinate various physiological processes in the body. Unlike the nervous system, which uses electrical impulses for rapid communication, the endocrine system utilizes **hormones**, which are chemical messengers released into the bloodstream to affect distant target organs. **Glands** are specialized organs that secrete hormones and are essential for maintaining homeostasis and regulating numerous bodily functions.

1. Overview of Hormones

Hormones are biochemical substances produced by endocrine glands that travel through the bloodstream to target tissues or organs, where they elicit specific physiological responses. Hormones can be categorized based on their chemical structure and function:

A. Chemical Classification of Hormones

Peptide Hormones

Structure: Composed of chains of amino acids.

Examples: Insulin, growth hormone, and antidiuretic hormone (ADH).

Mechanism: Bind to receptors on the surface of target cells, triggering a response through secondary messengers like cyclic AMP (cAMP).

Steroid Hormones

Structure: Derived from cholesterol.

Examples: Cortisol, estrogen, and testosterone.

Mechanism: Cross the cell membrane and bind to intracellular receptors, directly influencing gene expression.

Amino Acid Derivatives

Structure: Derived from single amino acids.

Examples: Thyroxine (from tyrosine) and adrenaline (from tyrosine).

Mechanism: Can act through surface receptors or directly enter the cell to influence cellular processes.

Eicosanoids

Structure: Derived from fatty acids.

Examples: Prostaglandins and leukotrienes.

Mechanism: Act locally (paracrine signaling) and influence processes like inflammation and blood clotting.

2. Major Endocrine Glands and Their Functions

Endocrine glands are specialized organs that produce and release hormones into the bloodstream. Each gland has specific functions and regulates different aspects of bodily function.

A. Pituitary Gland

The pituitary gland, often called the "master gland," is located at the base of the brain and controls other endocrine glands. It has two main lobes:

Anterior Pituitary (Adenohypophysis)

Hormones Secreted: Growth hormone (GH), thyroid-stimulating hormone (TSH), adrenocorticotropic hormone (ACTH), luteinizing hormone (LH), follicle-stimulating hormone (FSH), and prolactin.

Functions: Regulates growth, metabolism, stress responses, and reproductive processes.

Posterior Pituitary (Neurohypophysis)

Hormones Secreted: Antidiuretic hormone (ADH) and oxytocin.

Functions: ADH regulates water balance by influencing kidney function, while oxytocin stimulates uterine contractions during childbirth and milk ejection during lactation.

B. Thyroid Gland

Located in the neck, the thyroid gland produces hormones that regulate metabolism:

Thyroxine (T4) and Triiodothyronine (T3)

Function: Regulate metabolism, energy levels, and overall growth and development. T3 is the more active form of thyroid hormone, while T4 is converted into T3 in tissues.

Calcitonin

Function: Lowers blood calcium levels by promoting calcium deposition in bones.

C. Parathyroid Glands

The parathyroid glands are located on the posterior surface of the thyroid gland and secrete parathyroid hormone (PTH).

Parathyroid Hormone (PTH)

Function: Increases blood calcium levels by promoting calcium release from bones, absorption in the intestines, and reabsorption in the kidneys.

D. Adrenal Glands

Located on top of each kidney, the adrenal glands are divided into two regions:

Adrenal Cortex

Hormones Secreted: Cortisol, aldosterone, and sex hormones (androgens and estrogens).

Functions: Cortisol regulates metabolism and stress responses; aldosterone controls sodium and potassium balance; sex hormones influence secondary sexual characteristics.

Adrenal Medulla

Hormones Secreted: Adrenaline (epinephrine) and norepinephrine.

Functions: These hormones prepare the body for "fight-or-flight" responses by increasing heart rate, blood pressure, and glucose levels.

E. Pancreas

The pancreas functions as both an endocrine and exocrine gland.

Endocrine Functions

Hormones Secreted: Insulin and glucagon.

Functions: Insulin lowers blood glucose levels by promoting glucose uptake in cells, while glucagon increases blood glucose levels by stimulating glucose release from the liver.

Exocrine Functions

Function: Produces digestive enzymes that are released into the small intestine.

F. Gonads

The gonads (testes in males and ovaries in females) produce sex hormones and gametes.

Testes

Hormones Secreted: Testosterone.

Functions: Regulates sperm production and secondary sexual characteristics such as facial hair and deep voice.

Ovaries

Hormones Secreted: Estrogen and progesterone.

Functions: Regulate menstrual cycles, pregnancy, and secondary sexual characteristics such as breast development.

G. Thymus

Located behind the sternum, the thymus plays a crucial role in the immune system.

Hormones Secreted: Thymosins.

Functions: Stimulate the development and differentiation of T lymphocytes (T cells), which are essential for immune responses.

3. Hormonal Regulation and Feedback Mechanisms

Hormone levels in the bloodstream are tightly regulated through feedback mechanisms:

A. Negative Feedback

Negative feedback mechanisms help maintain homeostasis by reducing the output of a process when its effects reach a certain level. For example:

Thyroid Hormone Regulation: High levels of thyroid hormones (T3 and T4) inhibit the release of TSH from the pituitary gland, which in turn reduces thyroid hormone production.

Blood Glucose Regulation: High blood glucose levels stimulate insulin release, which lowers blood glucose. Low blood glucose levels reduce insulin secretion.

B. Positive Feedback

Positive feedback mechanisms amplify a process until an endpoint is reached. Examples include:

Childbirth: During labor, the release of oxytocin increases uterine contractions, which leads to more oxytocin release until childbirth occurs.

Lactation: Suckling stimulates the release of oxytocin, which promotes milk ejection and further stimulates suckling.

4. Disorders of the Endocrine System

Disorders in hormone production or regulation can lead to various health conditions:

Diabetes Mellitus

Type 1 Diabetes: An autoimmune condition where the pancreas produces little or no insulin.

Type 2 Diabetes: Characterized by insulin resistance and relative insulin deficiency.

Hyperthyroidism

Condition: Overproduction of thyroid hormones, leading to symptoms like weight loss, rapid heartbeat, and heat intolerance.

Hypothyroidism

Condition: Underproduction of thyroid hormones, causing fatigue, weight gain, and cold intolerance.

Addison's Disease

Condition: Adrenal insufficiency, leading to fatigue, weight loss, and low blood pressure.

Cushing's Syndrome

Condition: Overproduction of cortisol, causing symptoms like obesity, hypertension, and skin thinning.

Polycystic Ovary Syndrome (PCOS)

Condition: Hormonal imbalance in females that can lead to irregular menstrual cycles, excessive hair growth, and ovarian cysts.

The endocrine system, with its complex network of hormones and glands, plays a critical role in regulating various physiological processes. Understanding the functions of different hormones and glands, along with the mechanisms of hormonal regulation, provides insight into how the body maintains homeostasis and adapts to changes. Disorders of the endocrine system can have significant impacts on health and well-being, highlighting the importance of maintaining proper hormonal balance.

Fertilization and Embryonic Development

Reproduction and development are fundamental biological processes that ensure the continuation of species and the proper growth and development of offspring. This chapter covers the complex processes of **fertilization** and **embryonic development**, detailing how a single cell can develop into a fully formed organism.

1. Fertilization

Fertilization is the process by which a sperm cell from the male merges with an egg cell from the female to form a zygote, the first cell of a new organism. This process involves several stages:

A. Sperm and Egg Production

Spermatogenesis

Location: Occurs in the testes.

Process: Spermatogenesis produces spermatozoa (sperm cells) through meiosis. Each sperm cell is haploid, containing 23 chromosomes, which is crucial for maintaining the species-specific chromosome number after fertilization.

Oogenesis

Location: Occurs in the ovaries.

Process: Oogenesis produces oocytes (egg cells) through meiosis. Unlike sperm, females are born with a finite number of oocytes, which mature and are released during the menstrual cycle. Each egg cell is haploid, containing 23 chromosomes.

B. Stages of Fertilization

Sperm Capacitation

Definition: Capacitation is the process by which sperm undergo physiological changes in the female reproductive tract, making them capable of fertilizing an egg.

Changes: Includes the removal of glycoprotein coating on the sperm's surface and increased motility.

Acrosomal Reaction

Definition: The acrosomal reaction is the release of enzymes from the acrosome (a cap-like structure on the head of the sperm) that helps the sperm penetrate the egg's protective layers.

Process: These enzymes digest the zona pellucida, a glycoprotein layer surrounding the egg, allowing the sperm to reach the egg's plasma membrane.

Fusion of Sperm and Egg

Process: The sperm's plasma membrane fuses with the egg's plasma membrane, allowing the sperm's nucleus to enter the egg. This fusion triggers changes in the egg that prevent other sperm from entering.

Formation of the Zygote

Definition: Once the sperm's nucleus merges with the egg's nucleus, a diploid zygote is formed, containing 46 chromosomes (23 from each parent).

Process: The zygote undergoes its first mitotic division shortly after fertilization, marking the beginning of embryonic development.

2. Embryonic Development

Embryonic development involves a series of stages through which the zygote transforms into a fully developed fetus. This process occurs in the uterus and is divided into several key phases:

A. Early Development Stages

Cleavage

Definition: Cleavage is a series of rapid mitotic divisions of the zygote without significant growth between divisions.

Process: The zygote divides into smaller cells called blastomeres. This process results in the formation of a **morula**, a solid ball of cells, and then a **blastocyst**, which is a hollow structure with an inner cell mass.

Implantation

Definition: Implantation is the process by which the blastocyst embeds itself into the uterine lining (endometrium).

Process: The blastocyst secretes enzymes that help it penetrate the endometrium and establish a connection with the maternal blood supply.

Gastrulation

Definition: Gastrulation is a process during which the blastocyst reorganizes into a three-layered structure called the **gastrula**.

Layers:

Ectoderm: The outer layer, which develops into the skin and nervous system.

Mesoderm: The middle layer, which forms muscles, bones, and the circulatory system.

Endoderm: The inner layer, which gives rise to the digestive and respiratory systems.

B. Organogenesis

Definition: Organogenesis is the formation of organs from the three embryonic layers.

Process: Major organs and systems begin to develop, including the heart, brain, spinal cord, and other critical structures. This phase involves intricate signaling pathways and interactions between cells.

Key Developments:

Neural Tube Formation: The ectoderm forms the neural tube, which eventually develops into the brain and spinal cord.

Heart Development: The mesoderm forms the heart, which begins beating and pumping blood early in embryonic development.

Limbs and Organ Systems: Buds for limbs appear, and the basic structures of organs such as the liver, kidneys, and lungs develop.

C. Fetal Development

Transition to Fetal Stage

Definition: The transition from embryo to fetus occurs around the 8th week of pregnancy when all major organs and systems are in place.

Characteristics: The fetus undergoes significant growth and maturation of organ systems, preparing for survival outside the womb.

Fetal Growth and Maturation

Early Fetal Period (Weeks 9-12): Rapid growth of the fetus, development of facial features, and the start of movements.

Second Trimester (Weeks 13-26): Continued growth, development of more distinct features, and movements become more noticeable. The fetus begins to develop fat stores and practices breathing movements.

Third Trimester (Weeks 27-Birth): Final growth and maturation of the organs. The fetus increases in size, develops a more regular sleep-wake cycle, and prepares for birth.

D. Birth Process

Labor

Definition: Labor is the process of childbirth, involving the dilation of the cervix, contractions of the uterus, and the eventual delivery of the baby.

Stages:

First Stage: Includes early labor (cervical dilation begins), active labor (stronger contractions and full dilation), and transition (intense contractions and completion of dilation).

Second Stage: Involves pushing and delivery of the baby.

Third Stage: Delivery of the placenta.

Postnatal Period

Definition: The postnatal period encompasses the time after birth when the newborn adapts to life outside the womb.

Care: Includes monitoring the baby's health, feeding, and bonding with parents.

3. Genetic and Environmental Influences on Development

Genetic Factors

Role: Genes inherited from both parents influence the development of physical traits, susceptibility to certain diseases, and overall health.

Examples: Genetic disorders such as cystic fibrosis or Down syndrome are due to mutations or chromosomal abnormalities.

Environmental Factors

Role: Environmental influences, including maternal nutrition, exposure to toxins, and prenatal care, can affect fetal development and long-term health.

Examples: Adequate folic acid intake is crucial for preventing neural tube defects; exposure to teratogens (e.g., alcohol, drugs) can cause developmental abnormalities.

4. Common Developmental Disorders

Congenital Disorders

Examples: Heart defects, cleft lip/palate, and Down syndrome. These conditions are present at birth and can result from genetic or environmental factors.

Developmental Delays

Examples: Conditions like autism spectrum disorder (ASD) and intellectual disabilities. These may be identified during childhood and involve delays in physical, cognitive, or social development.

Fertilization and embryonic development are complex processes that transform a single fertilized egg into a fully developed organism. Understanding these processes provides insights into how genetic and environmental factors influence development and highlights the intricate mechanisms that ensure successful reproduction and growth. Awareness of potential disorders and developmental issues underscores the importance of prenatal care and early intervention for promoting healthy development.

Growth and Aging

The processes of growth and aging are fundamental aspects of biological life, involving the continuous changes that organisms undergo from conception to death. This chapter explores how organisms grow and develop throughout their lives and how they age over time.

1. Growth

Growth is the increase in size and mass of an organism over time, resulting from cellular proliferation and differentiation. It encompasses several stages and is influenced by genetic, environmental, and physiological factors.

A. Stages of Growth

Prenatal Growth

Zygote to Embryo: Following fertilization, the zygote undergoes cleavage and forms a blastocyst, which implants into the uterine wall. The embryo then undergoes gastrulation and organogenesis, leading to the development of major body systems.

Fetal Development: During the fetal stage, growth is characterized by the rapid increase in size and maturation of organs and tissues.

Infancy and Childhood

Infancy: From birth to about two years, rapid physical and neurological development occurs. Infants experience significant growth in weight and length, with major milestones including motor skills, sensory abilities, and cognitive development.

Childhood: The growth rate slows compared to infancy but continues steadily. During this period, children develop complex motor skills, cognitive abilities, and social behaviors.

Adolescence

Puberty: Adolescence is marked by the onset of puberty, which includes hormonal changes leading to sexual maturity and growth spurts. Boys and girls experience changes such as increased height, development of secondary sexual characteristics, and changes in body composition.

Physical Changes: Growth plates in long bones close at the end of adolescence, marking the end of height increase. However, body composition continues to change, and individuals reach their peak bone mass and muscle strength.

Adulthood

Early Adulthood: Growth is generally completed, but individuals continue to develop in terms of strength, endurance, and peak physical fitness. Psychological and cognitive development also continues.

Middle Adulthood: Growth is stable, but aging processes begin to manifest, such as gradual declines in muscle mass and bone density.

Late Adulthood

Senescence: Growth ceases, and aging processes become more pronounced. This period involves gradual physical and cognitive decline, including reductions in muscle mass, bone density, and organ function.

B. Factors Affecting Growth

Genetic Factors

Role: Genetics play a crucial role in determining growth patterns, including height, body composition, and susceptibility to certain diseases.

Examples: Genetic disorders such as Marfan syndrome or achondroplasia can impact growth and development.

Nutritional Factors

Role: Proper nutrition is essential for healthy growth and development. Deficiencies or excesses in nutrients can lead to growth disorders or health problems.

Examples: Adequate intake of protein, vitamins, and minerals supports normal growth. Malnutrition can lead to stunted growth and developmental delays.

Hormonal Factors

Role: Hormones regulate various aspects of growth. Growth hormone (GH), thyroid hormones, and sex hormones influence growth rates and development.

Examples: GH deficiency can result in growth disorders like dwarfism, while excess GH can lead to gigantism or acromegaly.

Environmental Factors

Role: Environmental conditions, such as exposure to toxins, stress, and overall living conditions, can impact growth and development.

Examples: Prenatal exposure to pollutants or maternal stress can affect fetal development and postnatal growth.

2. Aging

Aging, or senescence, refers to the gradual decline in biological function and efficiency over time. This process affects all organisms and leads to a progressive deterioration in physiological and cognitive functions.

A. Biological Theories of Aging

Genetic Theories

Programmed Theories: Suggest that aging follows a predetermined genetic program. For example, certain genes may regulate the lifespan and cellular repair mechanisms.

Examples: The telomere theory posits that telomeres (protective caps on chromosomes) shorten with each cell division, eventually leading to cellular senescence.

Damage Theories

Wear and Tear: Propose that aging results from accumulated damage to cells and tissues over time. This damage can be caused by environmental factors, metabolic byproducts, and genetic mutations.

Examples: Oxidative stress theory suggests that free radicals cause cellular damage, contributing to aging and age-related diseases.

Molecular and Cellular Theories

Protein Damage: Focus on the accumulation of damaged proteins and impaired cellular functions. Age-related changes in protein structure and function can lead to diseases like Alzheimer's.

Mitochondrial Dysfunction: Propose that aging is associated with declining mitochondrial function, impacting energy production and cellular health.

B. Physical Changes with Aging

Musculoskeletal System

Bone Density: Decreases with age, leading to conditions like osteoporosis and an increased risk of fractures.

Muscle Mass: Declines with age, resulting in sarcopenia (loss of muscle mass and strength).

Cardiovascular System

Heart Function: Heart muscle loses elasticity, and the efficiency of the cardiovascular system decreases. This can lead to conditions such as hypertension and heart disease.

Blood Vessels: Arteries become stiffer, increasing the risk of cardiovascular events.

Nervous System

Cognitive Function: Aging can lead to declines in memory, attention, and processing speed. Neurodegenerative diseases such as Alzheimer's and Parkinson's disease are more common in older adults.

Sensory Systems: Sensory abilities, including vision and hearing, often decline with age.

Immune System

Immune Function: The immune system becomes less effective with age, increasing susceptibility to infections and reducing the response to vaccinations.

Integumentary System

Skin Changes: The skin becomes thinner, less elastic, and more prone to damage. Wrinkles and age spots are common.

Hair Changes: Hair grays and thins over time.

C. Psychological and Social Aspects of Aging

Cognitive Changes

Memory: Short-term memory and processing speed may decline, but long-term memory and wisdom often remain intact.

Learning: Older adults may experience slower learning rates but can benefit from accumulated life experiences.

Emotional Well-being

Mental Health: Aging can impact mental health, leading to conditions such as depression or anxiety. Social support and engagement play crucial roles in maintaining emotional well-being.

Social Roles and Relationships

Retirement: Transitioning from work to retirement can affect identity and social interactions. Maintaining social connections and engaging in meaningful activities are important for quality of life.

End-of-Life Issues

Healthcare: Managing chronic illnesses and end-of-life care requires careful planning and support.

Palliative Care: Focuses on providing comfort and improving quality of life for individuals with serious, life-limiting conditions.

Growth and aging are complex processes that reflect the continuous changes in an organism's life. From the rapid development during infancy to the gradual decline in later years, these processes are influenced by a multitude of factors including genetics, environment, and lifestyle. Understanding the mechanisms underlying growth and aging provides insights into how organisms develop, how they can maintain health and well-being throughout their lives, and how they adapt to the challenges of aging.

Chapter 26: Conclusion
Glossary of Terms

In a comprehensive study of biology, understanding key terminology is crucial for grasping the concepts discussed. This chapter provides a glossary of important terms that have been covered throughout the various chapters of this textbook. Each term is defined in a manner that reflects its relevance to biological principles and processes.

A

Acid: A substance that donates hydrogen ions (H⁺) in a solution and has a pH less than 7. Acids can influence enzyme activity and biological processes.

Active Transport: The movement of molecules across a cell membrane from a region of lower concentration to a region of higher concentration, requiring energy input.

Adaptation: A trait or characteristic that enhances an organism's ability to survive and reproduce in a particular environment.

Aging: The process of gradual deterioration in biological functions and capabilities over time, leading to increased susceptibility to diseases and reduced physiological resilience.

Amino Acids: Organic compounds that serve as the building blocks of proteins. There are 20 different amino acids that combine in various sequences to form proteins.

B

Bacteria: Single-celled prokaryotic organisms that can be found in diverse environments. They play essential roles in processes such as nutrient cycling and are both beneficial and pathogenic.

Base: A substance that accepts hydrogen ions (H⁺) in a solution and has a pH greater than 7. Bases are crucial in maintaining pH balance in biological systems.

Biology: The scientific study of life and living organisms, encompassing their structure, function, growth, origin, evolution, and distribution.

Biosynthesis: The process by which living organisms produce complex molecules from simpler ones, often involving enzyme-catalyzed reactions.

C

Carbohydrates: Organic compounds composed of carbon, hydrogen, and oxygen that serve as a primary source of energy and structural components in organisms.

Cell Cycle: The series of events that a cell goes through as it grows and divides, including stages such as interphase, mitosis, and cytokinesis.

Cell Membrane: The phospholipid bilayer that surrounds and protects the cell, regulating the movement of substances in and out of the cell.

Cellular Respiration: The metabolic process by which cells convert glucose and oxygen into energy (ATP), carbon dioxide, and water.

Chromosome: A thread-like structure composed of DNA and proteins that carries genetic information. Humans have 23 pairs of chromosomes.

D

Diffusion: The passive movement of molecules from an area of higher concentration to an area of lower concentration until equilibrium is reached.

DNA (Deoxyribonucleic Acid): The molecule that carries genetic instructions for the development, functioning, growth, and reproduction of all known organisms.

Development: The process by which organisms grow and mature from a single cell to a fully developed organism.

E

Ecosystem: A biological community of interacting organisms and their physical environment, functioning as a system.

Enzyme: A protein that acts as a biological catalyst, speeding up chemical reactions in living organisms without being consumed in the process.

Evolution: The process through which populations of organisms change over time through variations in traits, natural selection, and genetic drift.

F

Fertilization: The process in which a sperm cell from a male merges with an egg cell from a female to form a zygote.

Fungi: A kingdom of eukaryotic organisms that includes yeasts, molds, and mushrooms. They play essential roles in decomposition and nutrient cycling.

G

Gene: A segment of DNA that contains instructions for synthesizing proteins or, in some cases, RNA molecules.

Genetic Engineering: The manipulation of an organism's genome using biotechnology to alter its genetic makeup.

Growth: The increase in size and mass of an organism or its parts over time.

H

Homeostasis: The process by which an organism maintains a stable internal environment despite changes in external conditions.

Hormones: Chemical messengers produced by glands in the endocrine system that regulate various physiological processes in the body.

I

Immune System: The complex network of cells, tissues, and organs that work together to defend the body against pathogens and diseases.

Inheritance: The process by which genetic traits are passed from parents to offspring through genes.

J

Juvenile: The stage of development following infancy and preceding adulthood, characterized by continued growth and maturation.

K

Kinetochore: A protein structure on the centromere of a chromosome where spindle fibers attach during cell division.

L

Lipid: A diverse group of hydrophobic molecules that include fats, oils, and phospholipids. They serve as energy storage, cell membrane components, and signaling molecules.

Mutation: A change in the DNA sequence of an organism's genome. Mutations can be beneficial, neutral, or harmful.

M

Meiosis: A type of cell division that reduces the chromosome number by half, producing four genetically unique gametes (sperm or egg cells).

Mitosis: The process of cell division that results in two daughter cells with the same number of chromosomes as the parent cell.

N

Natural Selection: The process by which organisms with advantageous traits are more likely to survive and reproduce, leading to evolutionary changes in populations.

Nucleus: The membrane-bound organelle in eukaryotic cells that contains the cell's genetic material (DNA).

O

Osmosis: The diffusion of water across a semipermeable membrane from an area of lower solute concentration to an area of higher solute concentration.

Organism: Any individual living entity that can carry out life processes independently.

P

Photosynthesis: The process by which green plants and some other organisms use sunlight to synthesize foods with the help of chlorophyll, converting carbon dioxide and water into glucose and oxygen.

Punnett Square: A tool used to predict the genotype and phenotype combinations in genetic crosses.

Q

Quantitative Data: Data that is expressed in numerical terms, often used in scientific research to measure variables and analyze results.

R

Reproduction: The biological process by which new individual organisms are produced from their parents.

Respiration: The metabolic process by which cells obtain energy from glucose and oxygen, releasing carbon dioxide and water as byproducts.

S

Speciation: The evolutionary process by which new biological species arise from existing species.

Symbiosis: The interaction between two different organisms living in close physical proximity, often to the benefit of both.

T

Tissue: A group of cells with a similar structure and function that work together to perform specific tasks in an organism.

Trait: A characteristic or feature of an organism that can be inherited or acquired, such as eye color or height.

U

Uptake: The process by which cells absorb nutrients, ions, or other substances from their environment.

V

Vaccine: A biological preparation that provides immunity against a specific disease by stimulating the body's immune response.

W

Watson and Crick: The scientists who first described the double-helix structure of DNA, which is fundamental to understanding genetics.

X

X-linked Traits: Traits that are associated with genes located on the X chromosome. They can exhibit different patterns of inheritance in males and females.

Y

Y Chromosome: One of the two sex chromosomes in humans, determining male sex characteristics and contributing to the development of male reproductive organs.

Z

Zygote: The initial cell formed when a sperm cell fertilizes an egg cell, which will develop into an embryo and eventually a full organism.

A thorough understanding of biological terminology is essential for comprehending the concepts and processes discussed throughout the study of biology. This glossary serves as a reference to reinforce key terms and ensure clarity in the discussion of complex biological topics. Mastery of these terms provides a foundation for further study and application of biological principles in various scientific and practical contexts.

Common Abbreviations in Biology

In biological sciences, abbreviations and acronyms are frequently used to simplify communication and notation. These abbreviations can represent everything from molecular structures to specific biological processes. Understanding these abbreviations is essential for interpreting scientific literature and engaging in discussions within the field. This chapter provides a comprehensive overview of common abbreviations used in biology.

A

ATP (Adenosine Triphosphate): The primary energy carrier in cells. It provides the energy needed for many cellular processes through the hydrolysis of its phosphate bonds.

ADP (Adenosine Diphosphate): A product of ATP hydrolysis; it can be converted back into ATP through cellular respiration.

AIDS (Acquired Immunodeficiency Syndrome): A disease caused by the Human Immunodeficiency Virus (HIV) that damages the immune system.

ARS (Adenosine Rich Sequence): A regulatory sequence in eukaryotic genes that can influence the stability and translation of mRNA.

B

bDNA (Big DNA): Refers to long DNA molecules or large DNA fragments used in genetic research.

BLAST (Basic Local Alignment Search Tool): A bioinformatics tool used for comparing an input sequence against a database to find similar sequences.

BSA (Bovine Serum Albumin): A protein derived from cows, commonly used as a standard in protein quantification assays and in various laboratory experiments.

C

cDNA (Complementary DNA): DNA synthesized from a messenger RNA (mRNA) template through reverse transcription, often used to clone genes or study gene expression.

CRISPR (Clustered Regularly Interspaced Short Palindromic Repeats): A revolutionary gene-editing technology that allows for precise modifications to DNA sequences.

CV (Coefficient of Variation): A statistical measure of the relative variability of a data set, often used in biology to assess the precision of experimental measurements.

D

DNA (Deoxyribonucleic Acid): The molecule that carries genetic information in all living organisms and many viruses.

DNP (2,4-Dinitrophenol): A chemical used in metabolic studies to measure the rate of respiration by uncoupling oxidative phosphorylation.

DTT (Dithiothreitol): A chemical reagent used to reduce disulfide bonds in proteins and stabilize proteins during biochemical experiments.

E

E. coli (Escherichia coli): A common bacterium used as a model organism in molecular biology and genetics research.

ELISA (Enzyme-Linked Immunosorbent Assay): A laboratory technique used to detect and quantify proteins, hormones, and other molecules using antigen-antibody interactions.

eGFP (Enhanced Green Fluorescent Protein): A fluorescent marker used to visualize and track gene expression and protein localization in living cells.

F

FACS (Fluorescence-Activated Cell Sorting): A technique used to separate and analyze cells based on their fluorescence characteristics.

FRET (Förster Resonance Energy Transfer): A technique used to study protein-protein interactions and molecular dynamics in live cells through fluorescence.

G

GMO (Genetically Modified Organism): An organism whose genome has been altered using genetic engineering techniques to express desirable traits.

PCR (Polymerase Chain Reaction): A technique used to amplify specific DNA sequences, enabling their analysis and manipulation.

GTP (Guanosine Triphosphate): A nucleotide involved in energy transfer within cells, similar to ATP, and important in signal transduction pathways.

H

HIV (Human Immunodeficiency Virus): The virus responsible for AIDS, which targets and destroys immune system cells.

HPLC (High-Performance Liquid Chromatography): A technique used to separate, identify, and quantify components in a liquid sample based on their interactions with a stationary phase.

HTS (High-Throughput Screening): A method used to quickly conduct millions of chemical, genetic, or pharmacological tests.

I

IP (Immunoprecipitation): A technique used to isolate and concentrate a specific protein or protein complex from a mixture using an antibody.

RNAi (RNA Interference): A biological process in which RNA molecules inhibit gene expression by destroying specific mRNA molecules.

IGF (Insulin-like Growth Factor): A group of proteins involved in growth and development, often studied in relation to growth hormone action.

J

JNK (c-Jun N-terminal Kinase): A protein kinase involved in stress responses and apoptosis signaling pathways.

K

KCl (Potassium Chloride): A chemical compound used in various biochemical and physiological experiments, including as a buffer or electrolyte.

kDa (Kilodalton): A unit of molecular weight used to express the size of proteins and other biomolecules.

L

LRP (Low-Density Lipoprotein Receptor-Related Protein): A receptor involved in lipid metabolism and cellular signaling.

LDH (Lactate Dehydrogenase): An enzyme involved in the conversion of lactate to pyruvate in cellular respiration.

M

mRNA (Messenger RNA): A type of RNA that carries genetic information from DNA to the ribosome, where it is used to synthesize proteins.

MHC (Major Histocompatibility Complex): A set of molecules displayed on cell surfaces that are involved in immune system recognition and response.

MT (Microtubules): Cytoskeletal components involved in maintaining cell shape, enabling cell movement, and facilitating intracellular transport.

N

NADH (Nicotinamide Adenine Dinucleotide): A coenzyme involved in redox reactions and energy production during cellular respiration.

NMR (Nuclear Magnetic Resonance): A technique used to determine the structure and dynamics of molecules, often used in structural biology.

NTP (Nucleoside Triphosphate): A molecule that provides the building blocks for nucleic acid synthesis and energy for various biological processes.

O

OSM (Osmotic Pressure): The pressure required to prevent the flow of water across a semipermeable membrane due to osmosis.

ORF (Open Reading Frame): A sequence of DNA that has the potential to be translated into a protein, starting with a start codon and ending with a stop codon.

P

pH: A scale used to measure the acidity or alkalinity of a solution. It reflects the concentration of hydrogen ions in the solution.

PCR (Polymerase Chain Reaction): A technique used to amplify specific DNA sequences, making them easier to study and analyze.

PLP (Pyridoxal Phosphate): The active form of vitamin B6, serving as a coenzyme in various enzymatic reactions.

Q

qPCR (Quantitative Polymerase Chain Reaction): A variation of PCR used to measure the quantity of a specific DNA or RNA sequence in a sample.

R

RNA (Ribonucleic Acid): A molecule that plays a central role in the expression of genetic information, including mRNA, tRNA, and rRNA.

RBC (Red Blood Cells): Cells in the blood that carry oxygen from the lungs to the body's tissues and transport carbon dioxide back to the lungs.

S

SNP (Single Nucleotide Polymorphism): A variation in a single nucleotide that occurs at a specific position in the genome among individuals.

SS (Single-Stranded): Refers to nucleic acids that consist of only one strand, such as single-stranded RNA (ssRNA) or DNA (ssDNA).

SDS-PAGE (Sodium Dodecyl Sulfate Polyacrylamide Gel Electrophoresis): A technique used to separate proteins based on their size and charge.

T

tRNA (Transfer RNA): A type of RNA that helps in the translation of mRNA into proteins by carrying specific amino acids to the ribosome.

TEM (Transmission Electron Microscopy): A microscopy technique used to visualize the internal structure of cells and tissues at high resolution.

U

UV (Ultraviolet): A type of electromagnetic radiation with wavelengths shorter than visible light, used in various biological applications such as DNA cross-linking and protein analysis.

Uptake: The process by which cells absorb substances, such as nutrients or ions, from their environment.

V

vDNA (Viral DNA): DNA that is found in viruses, which can be integrated into the host's genome or exist as an episome.

VEGF (Vascular Endothelial Growth Factor): A signal protein that stimulates the formation of blood vessels and is involved in processes like wound healing and tumor growth.

W

WBC (White Blood Cells): Cells of the immune system that are involved in defending the body against infections and foreign substances.

WT (Wild Type): The normal, non-mutated form of a gene or organism, used as a reference in genetic studies.

X

X-ray Crystallography: A technique used to determine the atomic and molecular structure of crystals by analyzing the pattern of X-ray diffraction.

Y

YAC (Yeast Artificial Chromosome): A vector used in genetic engineering to clone large DNA fragments in yeast cells.

Z

Z DNA: A left-handed DNA helix structure that contrasts with the more common right-handed B-DNA, involved in genetic regulation and expression.